Teachers: Learning to Bridge the Reading Gap

Teachers: Learning to Bridge the Reading Gap

Daniel Almeida

Good company makes the road seem shorter,
even on the wildest journeys, Horace

Table of Contents

Chapter 1 - Introduction

Background and Approach

This book presents an exploration of teachers' early reading professional learning within a complexity theory framework. It is an exploration that follows a research process for bringing teachers' voices to the forefront to understand and describe how contextual variables interact upon and influence teachers' professional learning, early reading knowledge and practice, and their students' early reading success. This research utilizes a qualitative multiple case study in a Québec school board with three schools chosen based on their demographic, social, and geographical differences and representative of the urban, suburban and rural characteristics of English schools and communities across the school board and the province. This study's structure opens up understandings of how professional learning opportunities are perceived by teachers to influence their instruction and students' reading development. Conceptualizing teaching and learning within a complexity theory lens, this research integrates iterative and deeply descriptive analysis to illuminate how contextual factors influence teacher learning and practice.

Problem Statement

The root of the problem motivating this study is the alarmingly high and consistent rate of students who fail to learn how to read proficiently and the associated effects of a range of adverse outcomes (Canadian Council on Learning, 2008; Canadian Literacy and Learning Network, 2009, 2012; Frontier College, 2017; Torgesen, 2000; Vanderstaay, 2006). In Canada, approximately 14% of 15-year old students were assessed as scoring below proficiency on the recent international PISA test with significantly higher achievement noted among those with greater socioeconomic advantage (O'Grady et al., 2019). Similarly, on the 2016 Pan-Canadian

Assessment Program, 11% of Grade 8 students in Québec scored below the baseline level

(O'Grady et al., 2018). In Toronto, Canada's largest city, its school board indicated that 53% of

students from the lowest income bracket and 34% from the highest income bracket were not

meeting provincial reading standards (Frontier College, 2017). The Canadian Council on

Learning (2008) project that by 2031, 36% of youth will have limited to very poor reading skills

making it hard to function in situations where reading is required and limit their chances in

adulthood for developing new skills for jobs that depend on skilled readers. Additional research

from the USA indicates that 60-70% of Grade 4 students read at a basic to below-basic level

(e.g., Allington, 2011; Washburn et al. 2011).

Noted above are the struggles that students from homes in the lowest income brackets

have for meeting provincial reading expectations. Another population of students who have

significant difficulties learning to read are students with learning disabilities who need specific

reading instruction and intervention in their early years to help decrease the chances of long term

reading failure (Wanzek et al., 2018). In 2019 the Ontario Human Rights Commission launched

Right to Read - a public inquiry based in the concern that students with reading disabilities were

not having their needs met in the public education system. The inquiry cited that many students

with special education needs (53% in 2018-2019) fail to score at or above the provincial average

in reading (EQAO, 2019). Longitudinal quantitative research interpreting how students with

reading disabilities in their early years are experiencing success later in life find students who do

not overcome reading disability and difficulty in their early years of school are less likely to

attain higher education and a similar income compared to their better reading peers (McLaughlin

et al., 2014). The range of students who struggle to learn how to read in school is large. This is

particularly so for students with specific characteristics; but it is prevalent for many students who are not considered to have a reading disability such as dyslexia or who come from homes not considered low on the socioeconomic scale (Allington, 2011; Canadian Council on Learning, 2010, Canadian Literacy and Learning Network, 2012). A static picture stretches back across years and decades that shows too many children are failing to learn to read at proficient levels and that failing to learn how to read more than likely negatively influences the lives of these children.

Connected to these alarming statistics and concerns, research affirms what is likely a general and widely held assumption: teachers are essential agents for impacting student achievement at school (e.g., Blazar & Kraft, 2017; Chetty et al., 2014; Rockoff, 2004). Primary grade teachers have the position and enormous responsibility to provide their students with their foundational academic and social skills, and often teachers are the ones criticized for not meeting their children's academic, physical and social-emotional needs (Allensworth et al., 2006; Cohen, 2006). Unfortunately, research indicates that many teachers do not possess a strong understanding of the core reading skills required to reach struggling readers and that teachers' negative perceptions towards assisting struggling readers, unfortunately, persist (Cunningham et al., 2004; Joshi et al., 2009; Moreau, 2014; Scharlach, 2008; Washburn et al., 2011).

Though we have evidence of effective early reading professional learning (e.g., Gersten et al., 2010; Joshi et al., 2009) and approaches to remediate struggling students' early reading achievement (e.g., Brady et al., 2009; Carlisle et al., 2011; McCutchen et al., 2002; Podhajski et al., 2009) it does not appear to translate into teacher practice and improved student reading outcomes. The problem is apparent, but the solution is more abstract. Too many students continue to fail to read proficiently. The solution involves a range of elements in need of an in-depth exploration: the role of individual teacher agency, the influence of varying contextual factors, and how the education system positions itself to prepare teachers for meeting the multifaceted and varied needs of students and teachers.

Osberg and Biesta (2008) suggest that "education purposely shapes the subjectivity of those being educated … [that] the function of education is to 'produce' certain kinds of subjectivities" (p. 314). These complexivists speak about the messiness of teachers' epistemologies, the composition of classrooms, and other varying factors at play within schools and influencing learning. A central point emphasized by Osberg and Biesta relates to the emergence of varying phenomena within the education system as being influenced by certain levels of academic achievement: "[E]ducation, for the most part (and for whatever sociopolitical reasons), is organized around the idea that its primary purpose is to promote 'outcomes' in certain areas deemed important (for whatever reason), and the curriculum (i.e. the content and pedagogy supporting it) is the primary tool by means of which such competency is achieved" (p. 315). Using this line of reasoning and considering teachers as learners within their systems of education is a unique way for considering acculturating influences upon teachers' learning, change and practices as they strive to help their students meet their curriculum's governed expectations. It is interesting to also take a further step outwards from the point of the teacher in the system. Consider the field of educational research and how a range of bodies of literature likely influence provincial planning documents, methods for implementing effective teaching practices, determining means and strategies for improving students' achievement, and shaping board and school priorities. Relatedly professional learning opportunities emerge out of these interacting perspectives connected in different degrees to overarching system expectations.

Purpose and Research Questions

In this research project, I seek to understand how teachers perceive the influences of their early reading professional learning on their practice and how they perceive that this shapes their students' reading development. To do this entails opening up and bringing together individual bodies of knowledge existing within the vast field of early reading teaching and learning and requires a research study directed towards understanding how early reading professional learning opportunities need to be more considerate of teachers' individual and contextual landscapes.

Research designed to understand contextual complexities influencing teacher learning and practice will help to unpack teachers' perceptions of the relationships between their early reading professional learning experiences, their classroom early reading practices and student reading achievement. Therefore, this research aims to open up intimate understandings and discussions of how their unique contexts influence teachers in three different schools in the same school board and how more universal aspects of teachers' early reading learning influence their learning and practice. Guiding my exploration are two open-ended questions:

1. How do contextual variables at the school, board, and provincial level influence the planning, delivery and uptake of early reading professional learning opportunities?

2. How do teachers perceive the relationships between (a) their professional learning experiences, (b) their classroom early reading practices, and (c) student reading outcomes?

Each research question helps to frame this exploration for understanding how teachers perceive that their early reading professional learning influences their early reading instruction, and subsequently, their students' reading outcomes. Of note, while I am choosing to use the term student reading outcomes in the second research question the importance is on the perception of teachers regarding students becoming confident and fluent readers versus any achievement measure.

Guided in a complexivist qualitative case study approach, these research questions are designed to draw out participants' perceptions in a way that can draw out a multifaceted approach to analysis and lead to within- and cross-case findings with theoretical, empirical and practical implications (Seidman, 2006, 2013, 2019).[1] The flow of this book emerges within a complexity-based perspective, placing value on understanding the influence of and upon human sensitivities to participants' contextually influenced interactions within and across systems. The next section brackets myself into the scope of this study.

Situating Myself

Initial interest in the topic of this research study emerged in my preservice (or Bachelor of Education) teacher education program fourteen years ago. As the first project in our problem-based learning cohort, my partner and I selected the state of current reading instruction to research. This assignment introduced us to the belief systems and reading practices described as a great debate by Jeanne Chall between proponents of phonics- and whole language-based early reading instruction. The tenets of this divide have stayed with me and appear influential and conjoined upon my path as teacher and curious researcher. My educational roots include being an elementary grade school teacher and a special education teacher who created individualized education plans for students with learning difficulties and exceptionalities. The complex nature of the school system, the perceived discord between research and theory, and a long-running

---

[1] Seidman's *Interviewing as qualitative research: A guide for researchers in education and the social sciences* is now in its 5th edition. Three editions of his text have been used in the course this book: The third edition (2006), 4th edition (2013) and the 5th edition (2019). Particular editions are referred to as they were used at different times through this research, though I predominantly used the 3rd edition as my go-to text on the 3-series interview process.

awareness of the number of students who struggle to learn to read have been a part of my life for the last 15 years. Therefore, this topic is not something appearing out of nowhere nor at minimum a shallow choice for the sake of a graduate research project. I am intimately concerned about what I have experienced as a teacher and as a researcher in elementary schools, and this research project helps unpack a perspective I feel is currently lacking in discussions on why too many students continue to struggle to learn to read.

As much as I seek to put personal and professional perspectives aside and equitably portray each participants' perspective, I am aware that aspects of my past will permeate elements of this research. I also ensure that I have maintained conscious integrity to the body of research, the participants' perspectives and present findings with theoretical, empirical and practical implications for early reading research and teacher learning. Underpinning my research method is a process for displaying each participant's perspectives and adherence to a qualitative research methodology to share the teachers' lived experiences (Seidman, 2006). Crafting teachers' lived experiences as profiles and vignettes in the first-person voice, their narratives provide a fertile starting point for an in-depth and saturated collection of findings to emerge. Iterative within-case findings are sensitive to local contexts. A transitional mode of analysis opens up an understanding of the variables and factors at play in early reading professional learning and instruction (Miles et al., 2014; Seidman, 2006), allowing richly representative cross-case findings across cases to emerge. This thorough process of analysis results in a conceptual framework that locates the individual teacher within the related contexts that the students they teach and the learning community that they learn and teach in share.

Rationale

Prominent reading researchers, social scientists and literacy networks relate the negative life experiences associated with remaining a poor reader (Allington, 2011; Canadian Council on Learning, 2008, 2010; Canadian Literacy and Learning Network, 2009, Castles et al., 2018; 2012 Snow et al., 1998; Torgesen, 2002). The importance of becoming a strong comprehender of written text is undeniable, and too many children struggle to learn how to read effectively. Renowned early reading researcher, J. K. Torgesen (2002), emphasizes that "[c]hildren who are poor readers at the end of first grade almost never acquire average-level reading skills by the end of elementary school" (p.8). Students with reading difficulties not resolved in their first years of school are more likely to incur teen pregnancies, have poorer health, drop out of high school, face incarceration and therefore are likely to be on the fringes of our society that values and depends on skills from literate citizens (Canadian Council on Learning, 2008, 2010; Canadian Literacy and Learning Network, 2009, 2012; Claessens et al., 2009; Frontier College, 2017; Morgan et al., 2008; Torgesen, 2002; Vanderstaay, 2006).

Thankfully, it should appear, that over the last two decades much research espousing evidence-based early reading instructional practices has been produced (e.g., Canadian Education Statistics Council, 2009; Canadian Literacy and Learning Network, 2009, 2012; National Reading Panel, 2000; Ontario, 2003; Snow et al., 1998, Torgesen, 2002). Awareness of these findings should lead to the assumption that there has been a significant decrease in the number of students not becoming proficient readers early on. Unfortunately reading research indicates that many teachers do not possess a core breadth of early reading knowledge, that often professional learning experiences do not effectively help them meet the needs of students who struggle to read, and that negative teacher perceptions towards struggling readers persist

(Cunningham et al., 2004; Joshi et al., 2009; Scharlach, 2008; Washburn et al., 2011).

Significance of the Study

Together the school (within-case) and the board (cross-case) findings provide an

inclusive interpretation of how teachers' perspectives are shaped and influenced. Understanding

perspectives in such a fashion provides a depth and richness that though unique to a specific

board within a specific region of a specific province, could be considered to reverberate to

alternative environments. This study illuminates how professional learning can be

conceptualized and understood through empathetic consideration of teachers' readiness and

needs, intimately connected to their students' early reading needs. Clarifying the challenges and

complexities that exist within boards, yet also shining a light on the similarities, opens avenues

for considering how early reading professional learning can be conceptualized for teachers'

preservice and inservice professional learning and that positively influence students' reading

achievement. This research opens up a way for considering more contextually relevant early

reading professional learning that meets teachers' professional needs.

The Roadmap

Chapter 1 broadly outlines this doctoral research project's scope, introducing the problem

and research questions and positioning the nature of my complexivist thinking and who I am as

an educator and educational researcher. Chapter 2 is multi-purpose: First is a brief background of

the state of early reading instruction in the province of Québec and beyond; second, a review of

the literature on approaches to early reading instruction, teacher early reading knowledge and

professional learning follows; and third, this chapter concludes by opening up complexity theory

as the conceptual framework guiding this study. Chapter 3 presents the methodology and frames the qualitative case study research methods within the complexity theory approach. This chapter provides a brief introduction of the participants, the unique and in-depth process of three-series interviews with teacher participants, the semi-structured interviews with principals and board-level reading experts, and the steps I undertook throughout data analysis. Chapter 4 comprises the Findings: Contextual introduction of the school board and the case schools, within-case themes nested in three within-case narratives, contextual networks representing pathways involving teachers' perceptions of how contextual variables are perceived to interact and influence their learning and their students' reading development, and finally, themes presented in a cross-case analysis that brings abstractions across cases together and provides a foundation to conceptualize teacher learning and student achievement. Chapter 5 transitions to a discussion on the findings, and Chapter 6 concludes this book by suggesting theoretical, empirical and practical implications to consider for how improving how professional learning can influence students' early reading needs.

Chapter 2 – The Contextual Landscape, Literature Review and Conceptual Framework

This chapter begins by providing a sense of the contextual landscapes of early reading instruction in Québec and North America. First is a brief overview of the Québec curriculum (the Québec Education Program [QEP]) and the state of early reading instruction in Québec. Second, is a portrayal of the landscape of the state of early reading and the swings in differing approaches to early reading instruction. Following this is the literature review on approaches to early reading instruction, teacher knowledge and beliefs, and professional learning. This chapter ends with an in-depth presentation of the role of complexity theory as the conceptual framework for this research design.

Context of the Québec Education Program and Language Arts Curriculum

Officially implemented in 2001, the preparations for Québec's current provincial education plan arose in the mid-1990s, (CEA, 2017). Designed to rewrite an outdated curriculum to be more in line with ideas for a 21st-century knowledge and technology-based society, "[t]he objective of the change was to refocus the school on its main mission: to educate, socialize and qualify today's youth … to return to the essential knowledge to be passed on, to raise the cultural level of programs, to avoid the compartmentalization of knowledge and to introduce more rigorous evaluation" (Guimond, 2009). The QEP curriculum is competency-based with "[k]nowledges … organized in terms of competencies to make learning meaningful and open-ended for students" (Gouvernement du Québec, 2001, p. 4). In the elementary years, grades are organized around three 2-year cycles (Cycle 1 comprises grades 1 and 2; Cycle 2, grades 3 and 4; and Cycle 3, grades 5 and 6) with the intention that this structure provides room for teachers to connect with students over two years and that teachers will be more able to differentiate instruction according to individual rates of student learning. The framework of the education program has three central organizing structures: cross-curricular competencies, broad areas of learning, and core subject competencies. Cross-curricular competencies "transcend the limits of subject-specific knowledges while they reinforce their application and transfer to concrete life situations precisely because of their cross-curricular nature … they are mutually complimentary" (p. 12) and organized into four categories. Five broad areas of learning are designed and intended to "help students relate subject-specific knowledges to their daily concerns and thus give them a better grasp of reality" (p. 42). See Table 1 to get a clearer picture of each of these three components and to consider the expectations teachers are responsible for across grades and cycles.

Table 1

Elementary Curriculum Expectations for Teachers in Québec

Cross-curricular Competencies	Broad Areas of Learning	Core Subject Areas
I. Intellectual Competencies i. To use information ii. To solve problems **II. To exercise critical judgement** iii. To use creativity **III. Methodological Competencies** iv. To adopt effective work methods v. To use information and communications technologies **IV. Personal and Social Competencies** vi. To construct his/her identity vii. To cooperate with others **V. Communication-related competency** viii. To communicate appropriately	**I. Health and Well-Being** **II. Personal and Career Planning** **III. Emotional Awareness and Consumer Rights and Responsibilities** **IV. Media Literacy** **V. Citizenship and Community Life**	**I. Languages** i. English Language Arts ii. French Language Arts **II. Mathematics, Science and Technology** **IV. Arts Education** **V. Personal Development**

For each subject area is a range of competencies (competencies considered as the broad behaviours, skills and knowledge and ability to communicate understanding students are to develop within each of the subjects), which act as the framework for learning. Competencies for each subject are the same across grades 1 to 6, and each competency breaks down into key features that are also the same across the grade levels. Key features are the processes comprising the competencies. Finally, the curriculum prescribes essential knowledges that students need to learn by the end of cycles in order to work towards mastery of competencies by the end of Grade 6. For example, the English Language Arts curriculum is organized first by the competencies, next the key features, and then the essential knowledge to be developed and worked on throughout the elementary years.

In 2009 the government released a document called the Progression of Learning to compliment the 2001 QEP and to "provide more information to teachers about some of the requirements found in the content of the ELA program and their connection to the progressive development of literacy from the beginning to the end of elementary school" (Gouvernement du Québec, 2009, p. 4). It provides more specific information about what expectations to teach in grades instead of just across two-year cycles. Laid out across the years under the umbrella of cycles, the Progression of Learning applies a symbol system to designate whether students are constructing knowledge, applying knowledge, or reinvesting in knowledge. The Progression of Learning provides a sense of at what grade students should be applying specific essential knowledges in terms of learning to read. However, it does not expand on the 2001 curriculum for specifying expectations for how students are to progress in learning how to read.

The English Language Arts Curriculum

In the Language Arts Curriculum, there are four subject level competencies that occur across the elementary years. These are: (a) to read and listen to literary, popular and information-based texts; (b) to write self-expressive, narrative and information-based texts, (c) to represent literacy in different media; and (d) to use language to communicate and learn (QEP, 2001).

The curriculum states expectations for teachers in Grade 1 and 2 to focus on similarly stated essential knowledges with their students across Cycle 1. In terms of learning to read and the expectations for teachers to help students learn the relationship between sound and print and to develop reading comprehension, the QEP provides broad guidelines, but nothing overtly specific relating to the early years. There is a particular reference in a section on essential knowledge of reading strategies for students to construct meaning from text by using the four cuing systems (semantic, pragmatic, syntax and graphophonemic). However, other than a relatively broad and passing reference, nothing explicitly guides teachers or provides information on what early reading teachers should concretely focus on to help their students learn to read (see Figure 1 below).

Figure 1

Essential Knowledges, Reading Systems, the Four Cuing Systems (QEP, p. 77)

READING STRATEGIES

The student uses the following repertoire of strategies to construct meaning from texts:

• The four cuing systems, which include:			
– Prior knowledge and personal experience of the content of a text (semantic)	❶	❷	❸
– Knowledge of the ways books work (pragmatic), e.g. most fairy tales begin with, "Once upon a time…"	❶	❷	❸
– Use of pictures and other graphic representations to interpret texts (pragmatic). See also Competency 3, re: reading texts that have images and illustrations	❶	❷	❸
– Knowledge of common language patterns (syntax). See also Competency 2, Writing System	❶	❷	❸
– Knowledge of the relationships between sounds and written symbols (graphophonics)	❶	❷	❸

Several other reading strategies relate to meaning-making, locating information and responding to different genres of text. However, Figure 1 reflects the depth of teaching early reading skills and strategies in Québec's English language curriculum. It appears that teachers, schools and boards are responsible for developing instructional strategies to meet the needs of their beginning readers.

English Language Arts in the Case School Board. This very brief section provides a general overview of the approach to English Language Arts that the school board in which this research project takes place. Reflecting the gist of the description of the English Language Arts curriculum in Québec, the case school board's English Language Arts program follows a Balanced Literacy Model that is broadly defined by the board as a model offering students time to read, time to converse and time to write.[2] It appears to be a holistic-type model and approach that provide students access to a wide range of literacy skills. Assessment of progression in the case school board comprises diagnostic reading assessments based on reading fluency norms at particular grade levels.

The Contextual Landscape of Reading Instruction

This brief section moves from the expectations of the provincial English Language Arts Curriculum to provide a sense of the field of early reading instruction and the reading wars and literacy paradigms occurring in North America for many decades or as Castles et al. (2018) point out, even centuries. For a more comprehensive background about the landscape of early reading, see Beach and O'Brien (2015), Castles et al. (2018), Kim (2008) and Perry (2012).

---

[2] I have withheld the link and reference for this information to maintain anonymity for the participating school board.

In general, the reading wars are coined as a debate between opposing philosophies to teaching reading; through phonics (or code-based approach) or a whole reading approach (Castles et al., 2018; Kim, 2008). Castles et al. (2018) citing a seminal work about the history of learning to read by Adams (1990) notes that the reading wars can be traced back approximately 200 years when the state of Massachusetts's Secretary of Education considered reading instruction teaching the relationships between sounds and letters to be a soulless enterprise. More recent are the swings in approaches to reading instruction: code-based-vs-whole language. In the 1960s, Jeanne Chall reported in her seminal work on reading instruction that instruction applying a code-based approach was more effective than teaching whole words and sentences; however, Kim (2008) pointed out that this was quickly reproached in research by Kenneth Goodman who argued that "reading was a 'psycholinguistic guessing game'" (Goodman, 1967, p.127), where good readers use context clues and their background knowledge "to predict, confirm, and guess at the identification of new words" (Kim, 2008, p. 373). To many, Goodman's work is the research foundation of the whole language movement in which reading is a natural process that requires immersion within an engaging literacy environment (Kim, 2008), though Goodman slyly disputes this (Goodman, 2014). Kim notes that between the 1970s and 1990s, research in the field of cognitive psychology uncovered the processes that good readers use. These findings comprise a synthesis of phonics and whole language approaches to reading and stress the importance of ensuring phonics instruction for students in their early years of learning to read. Castles et al. (2018) also note how more recently a balanced approach, or a three-cueing approach, that involves teaching semantic (using context, background knowledge to identify words and figure out meaning), syntactic (using the rules of word and sentence structure), and graphophonic (using the letter-sound, sound-symbol relationships) skills is what the current

reading research demonstrates is crucial for reading instruction to be most effective.

Interestingly, noted in the discussion on the Québec English Language Arts curriculum,

teachers are expected to teach students to use a 4-cuing system (QEP, 2001). In addition to the

semantic, syntactic and graphophonic in the prior sentence is a pragmatic aspect (using the social

and cultural aspects of language) to the cuing system. This simplified version of the rocky

contextual landscape of reading instruction presents a sense of the atmosphere in which students

and teachers work and learn. Though some of the literature review research will appear dated, it

is the essence of how teaching early reading instruction has generally occurred over the last four

decades and provides a sense of the research base examining the effect of three main approaches

to early reading instruction. Before moving to the literature review, the next section positions

literacy instruction within existing literacy paradigms from over the past century.

Literacy Paradigms

Beach and O'Brien (2017) present a historical review of influential paradigms of literacy

research since 1920. The choice for the research they review is based on how influential it

appeared to be on re-defining literacy learning perspectives and changes in English language arts

instruction and comprises four main periods. Formalist and behaviourist perspectives on literacy

learning were influential from 1920 to 1960. A formalist paradigm emphasizes the structuring of

language in text and writing and reading instruction that focuses on analyzing what the structures

of exemplary text are. The 1960s through to 1970 represents an increased focus on cognitive

research processes for understanding how to read and write. Emerging in these years were

psychological studies of fundamental processes in reading and writing development. Notably,

Beach and O'Brien remark that this period informs many of the current strategies for teaching

17

word reading (e.g., decoding) and comprehension instruction. In the cognitive paradigm, reading and writing are universal processes that anyone can learn from instruction on reducing elements of words or sentences and larger strings into essential small parts. Dillon and O'Brien (2019) argue for a pragmatic approach to reading research and instruction. They note the prevalence of cognitive processes privileged in scientific paradigms favouring the clear and measurable over ambiguities in sociocultural and qualitative research. For these researchers, this is problematic. It creates a culture prioritizing a simplistic style of thinking and research and maintains a one-sided form of reading instruction creating a deficit discourse and ignoring the effect of society, culture and context on learning to read for students who are not in a majority position in society.

The 1970s through the 1980s began drawing from Dewey's experiential progressivist learning theories with researchers adopting more sociocultural perspectives to focus on how participation in different social and cultural contexts shape literacy practices (Beach & O'Brien, 2017). Branching from this sociocultural paradigm is a notion of social and cultural literacy that challenges the traditional definitions of literacy privileging the dominant culture within their institutions. Sociocultural perspectives emphasize the social and cultural contexts literacy is practiced within and the interacting power relations at play in how literacies are defined. These perspectives differ, however, in how literacy is defined (meaning from print, meaning from and across multimodalities and meaning from reading the word and the world) (Perry, 2012). Perry describes sociocultural perspectives as:

> concerned with understanding how people use literacy in their everyday lives, finding
>
> ways to make literacy instruction meaningful and relevant by recognizing and
>
> incorporating students' out-of-school ways of practicing literacy, and decreasing

achievement gaps for students whose families and communities practice literacy in ways

that differ from those in the mainstream or in positions of power. (p. 51)

O'Brien and Rogers (2015) remark that sociocultural perspectives and frameworks have been a

part of literacy research and practice over the past four decades. However, though sociocultural

perspectives (e.g., literacy as a social practice, multiliteracies and critical literacy studies) are a

regular presence in reading research and instruction, they are not equal with cognitive and

psycholinguistic literacy paradigms which continue to dominate institutional expectations for

how teachers are to teach reading (Beach & O'Brien, 2017; Perry, 2012). Therefore, though there

is an increase in sociocultural literacy research, this is not infiltrating how many schools view

and teach reading (Dillon & O'Brien, 2019; Perry, 2012). Subsequently, though elements of the

following literature review represent traces of sociocultural perspectives, most of the research

comprises cognitive, psycholinguistic reading instruction paradigms.

Literature Review

This literature review, (a) opens up the interconnected landscape of early reading

instruction and professional learning through a critique of the extant scholarship in this area, and

(b) through this critique identifies under-explored areas in early reading teaching and learning. A

complexity theory perspective guides the intention and design of the literature review. Influential

to the shape of this literature review is Opfer and Pedder's (2011) complexity theory synthesis of

change literature, which brought together varying professional learning and teacher change

findings. In this scope, this review unpacks a range of literature connected to early reading

instruction: approaches in providing early reading instruction, teacher knowledge and beliefs,

and early reading professional learning designs. First, the literature on approaches to early

reading instruction is reviewed. This is followed by a review of the literature on teachers' perceptions of their role in early reading instruction. It concludes with a synthesis of the

literature on teacher early reading knowledge and the relevant teacher change and early reading

professional learning research. Emerging from this comprehensive review of the literature the

gaps in the field of early reading instruction and professional learning research are identified,

briefly discussed and positioned into the scope and direction of my research project.

Early Reading Instruction

This section presents a literature review of the three main approaches to early reading

instruction over the last half-century: code-based, whole language and balanced reading.

Code-based and Whole Language Approaches to Reading

'Code-based' is the term used to represent the approach to early reading instruction for

developing fluent reading through stages in which learners gradually acquire numerous subskills,

reading ability is mastered and eventually perceived ingrained (McKenna et al., 1990).[3]

Proponents of code-based early reading instruction generally understand beginning reading as a

step-by-step process that involves explicit phonemic awareness (ability to hear and manipulate

sounds in spoken words) and phonics instruction (ability to discriminate printed text and produce

its corresponding sound) for ensuring that print-sound correspondences happen in specific

sequences (Adlof et al., 2010).[4] Whole language reading approaches stress dependence on a

climate in which students exposed to enriching literacy environments will choose to read because

they want to read and the process of learning to read within this context echoes the natural way

---

[3] At times code-based instruction will be used interchangeably with phonics-based, skills-based or
traditional instruction due to research reported in those terms, though all terms infer the same meaning as described
in this paragraph.

[4] A **phoneme** is the smallest unit of a speech sound which can change the meaning of a word: /k/ is the
phoneme represented by the letter 'k' in the word kiss, and **phonemic awareness** is the ability to reflect on and
manipulate the phonemes in spoken language (Cain, 2010, p. 250).

children acquire language (Edelsky, 1990; Gee, 1995; Goodman, 1989). Elements considered parts of a whole language approach to reading are broad. They can include quality children's literature, daily read alouds, structured independent reading activities embedded throughout the curriculum and emphasizing higher-order thinking, collaborative groupings, teacher-student conferences, and grammar and spelling (Daniels et al., 1999).

A Comparison of the Whole Language Reading Approach to Code-based Instruction

Five meta-analyses compare whole language reading instruction to code-based instruction. In total, these meta-analyses include 237 research studies and 5 American education projects. These meta-analyses, published between 1989 and 2016, focus on bringing together and comparing the impact of early reading instructional approaches on various reading-related outcomes for average-achieving and struggling students. Outcomes considered in these meta-analyses measured reading readiness, beginning reading achievement and effective approaches and interventions, and student attitudes towards reading. Stahl and Miller (1989) conducted a meta-analysis comparing code-based instruction to whole language-based reading instruction, and Stahl et al. (1994) updated this analysis five years later. Gee (1995), concerned with Stahl and Miller's (1989) and Stahl et al.'s (1994) inclusion of the language experience approach, conducted a meta-analysis comparing code-based instruction to what he considers instruction more fully representative of whole language approaches. Suggate (2010, 2016) conducted two meta-analyses on the effect of reading approaches for students with reading disabilities, learning disabilities or at risk for struggling in reading.

Stahl and Miller (1989) and Stahl et al. (1994) use two procedures to analyze their data. The first was vote-counting, where each study classifies as either favouring whole language or basal reader approach with measures including standard achievement tests, attitude measures, miscue analysis and print concepts.[5] The second procedure is the same that Gee (1995) employs: effect sizes derived from the quantitative data looking at the mean differences resulting from the whole language reading group compared to the skills-based instruction. All three of these meta-analyses point to the whole language reading approaches providing slightly higher achievement in reading for kindergarten students. Stahl and Miller (1989) indicate that 17 kindergarten comparison studies favour whole language/language experience approaches, and two favour the basal reading. Gee's (1995) meta-analysis favoured the whole language approach for kindergarten to third-grade students demonstrating higher achievement scores than code-based approaches with an overall effect size of .70, though he cautions that sample sizes were small. Finally, when comparing students with and without learning disabilities, Gee found that whole language instruction has a positive influence on students' early reading (normal effect size =.59 and learning disability effect size =.53) but cautions that though effect sizes indicate whole reading's effectiveness these differences are not statistically significant between these groups. Gee also suggests that with larger sample sizes, the results could indicate that whole language is not as effective as initially considered.

Suggate's (2010) meta-analysis of 85 experimental or quasi-experimental studies investigates the effectiveness of reading intervention approaches for students from preschool through Grade 7 considered at-risk for reading difficulties or struggling to read. Broadly Suggate finds that phonemic awareness and phonics interventions are more effective in the early years and that whole language interventions demonstrate greater effect sizes in the later primary and elementary years. The reading interventions comprised phonics, phonemic awareness and whole

---

[5] Generally, basal reading programs equate to published reading programs where teachers teach students the basics of learning to read in a step-wise, generally code-based prescribed fashion (Smith et al., 2001).

language interventions for 116 treatment-control groups (N = 7,522). The following class levels represent the comparison groups: (a) preschool and kindergarten, (b) Grade 1, (c) Grade 2, (d) Grades 3 and 4, and (e) Grades 5, 6 and 7. Comparison groupings also comprised groups of at-risk and struggling readers. Effect sizes were generated based on the intervention (phonemic and phonological awareness, phonics, and whole language) on the different groupings of students.[6] Generally, the results indicate that there are statistically larger effect sizes, and therefore advantages to providing phonics (d =.59) and phonemic awareness (d =.67) for developing pre-reading skills that are foundational to reading and comprehension skills in the early grades (Pre-K to Grade 3). Whole language (comprehension-based) interventions had large effects (d =.69) on measures of reading (e.g., word reading, passage reading) and on comprehension (d =.60). Even though this meta-analysis looks only at intervention effectiveness on struggling readers across Pre-K to Grade 7 and comprises small instructor to student ratios (e.g., M = 3.68), it has relevance to my research. Suggate demonstrates that teaching students how to break the code (phonemic awareness, phonological awareness and phonics instruction) will help students who struggle to read in their critical early years and that providing an engaging whole language environment is critical in all elementary grades.

Suggate's (2016) meta-analysis explores the long term effects of reading interventions for average and struggling readers. The main differences to his 2010 research include changes to the number of intervention types, risk status (including average readers), more intervention features and outcome variables (e.g., the inclusion of spelling), and seeking to understand what interventions have long term effects approximately 11-12 months following posttest. Grouping

---

[6] **Phonological awareness** is the ability to reflect on and manipulate the spoken sounds in the language (Cain, 2010, p. 250).

levels also slightly differ: (a) Preschool and kindergarten, (b) Grades 1 and 2, and (c) Grades 3 to 6. The four categories of risk status are average, at-risk (low SES, reading below the 50th percentile, a parent with dyslexia), low readers (reading between the 11th and 25th percentiles), and reading and learning disabled (reading below the 10th percentile, having an IQ-reading discrepancy of 1 standard deviation, or a general learning disability). Interestingly average readers lost some advantage over control groups in follow-up measures. The largest effect-sizes at follow-up are phonemic and phonological awareness (posttest d =.42, follow-up d =.46) and comprehension interventions (posttest d =.33, follow-up d =.42). Phonics (posttest d =.44, follow-up d =.25) and fluency interventions (posttest d =.59, follow-up d =.33) indicate decreased effects at the follow-up. Of interest to my current research, the most effective interventions at posttest for preschool to Grade 2 students were phonemic and phonological awareness and phonics interventions. However, when compared at follow-up, the phonemic and phonological awareness intervention demonstrates a distinct advantage. Possible reasons are that phonics intervention groups had much higher sample sizes and were assigned greater weights that led to overall smaller effect sizes.

Focusing specifically on struggling early readers, research from Maddox and Feng (2013) suggest that a phonics-based approach is more impactful than a whole language approach for increasing spelling and word recognition ability. Maddox and Feng's (2013) quasi-experimental action research design includes 22 first grade students (13 boys, 9 girls) with students randomly assigned to either a whole language or phonics group with stratified sampling used. Out of the 12 students in each group 3 were assessed as above, at, or below grade level and over 4 weeks students met with their teacher each school day for 20 minutes of phonics or whole language reading instruction. Results indicate that though there was not a significant difference between the two groups the phonics groups showed greater improvements in reading and spelling and that both instructional methods related to noticeable changes in students' reading fluency (students in the phonics group improved reading fluency by 8 points, and students in the whole language group increased by approximately 4 points). Ryder et al. (2007) explore whether explicit instruction in phonemic awareness and phonemically-based decoding skills is an effective intervention for children with early reading difficulties taught in a mainly whole language instructional environment. The 24 lowest performing year 2 and 3 students in reading were organized into 12 closely matched pairs randomly assigned to an intervention (phonics) or control (whole language) group. The intervention occurred over 24 weeks during the first three terms of a four-term school year. Each group of three students received four lessons per week, varying between 20 and 30 minutes. The intervention was supplemental to their classroom whole language reading instruction. Posttest measures reveal the intervention group outperforming the control group at posttest on all measures of phonemic awareness, phonological decoding ability, the accuracy of recognizing words in connected text, and reading comprehension. In their two-year follow-up, Ryder et al. demonstrate that the phonics intervention group significantly outperforming the whole language control group.

Summary to code-based and whole language reading approaches. Three meta-analyses conducted between 1989 and 1995 conclude that both approaches to reading are comparable in their impact on students' reading achievement, but whole reading instruction appears to be more effective in kindergarten. Two meta-analyses conducted in 2010 and 2016 indicate that phonemic and phonological awareness and phonics interventions are effective approaches for helping struggling readers develop decoding skills and that phonemic and phonological interventions lead to large follow-up effects up to one year later (Suggate 2010,

2016). Implications of Suggate's (2010, 2016) findings indicate the value of providing students effective phonemic and phonological instruction in the early years, that phonics instruction leads to moderate to large effect sizes following intervention and that regular progress monitoring of pre-reading and reading skills is essential to ensure students maintain levels of pre-reading and reading skills and that they continue to develop. Results from Suggate's (2010, 2016) meta-analyses also highlight the effect of providing comprehension skills for early readers and whole language comprehension instruction across the elementary years for all students, regardless of reading ability. This contrasts in degrees with Stahl and Miller (1989) and Stahl et al. (1994) who indicate that whole language instruction in kindergarten might better prepare students by "promot[ing] children's conceptual base for reading" (Stahl et al., 1994, p. 176), but does support their findings that code-based instruction can be more effective in grade 1. As students reach grade 1 the research suggests that a more structured approach focusing on code-based learning leads to improved reading achievement (Maddox & Fengh, 2013; Ryder et al., 2007; Stahl & Miller, 1989; Stahl et al., 1994; Suggate 2010, 2016), but this can also be contested (Gee, 1995). Apparent above is research supporting the blending of code-based instruction with whole language-based approaches in boosting the long-term reading achievement of students with reading difficulties (Ryder et al., 2007; Suggate, 2010, 2016). A blend of code-based and whole language approaches can loosely be understood as the tenet of balanced reading instruction (see Freppon & Dahl, 1998) and is reviewed next.

Balanced Reading Instruction

A small body of research explores the impact of balanced reading approaches on student reading outcomes (e.g. Bitter et al., 2009; Diamond & Ongwuebuzie, 2000; Donat, 2006; Guthrie et al., 2001). Large-scale studies by Bitter et al., Donat, and Guthrie et al. find that students in balanced reading classes make significant gains and overall, the need for supplemental reading intervention decreases. More specifically, students receiving balanced reading instruction have higher reading comprehension and word recognition scores and demonstrate greater reading readiness than students not provided a balanced reading approach (Bitter et al.; Donat; Guthrie et al.). In contrast, Diamond and Ongwuebuzie find that balanced reading is associated with significantly lower reading achievement and poorer attitudes toward reading.

Of the four studies reviewed, three involving approximately 5400 students provide evidence that a balanced reading approach cultivates a readiness to read; and improves basic word reading ability and overall reading achievement (Bitter et al., 2009; Donat, 2006; Guthrie et al., 2001). Providing support that balanced instruction improves reading comprehension is research from Bitter et al. (2009). They explore a hypothesis that aligning instructional approaches within a balanced literacy model will improve student reading outcomes. Their intensive case study conducted in 101 classrooms from 9 primarily high poverty school districts that grade 3, 4 and 5 students' reading found that comprehension scores were positively related to balanced instruction elements.[7] Comparative research from Donat (2006) and Guthrie et al. (2001) provide robust experimental data supporting a balanced reading approach, notably in how it can effectively support struggling readers.

Donat (2006) evaluates the effectiveness of a balanced reading program implemented gradually into 12 schools to 4500 students. Overall findings indicate that reading readiness and word recognition achievement to be higher for students who receive balanced reading instruction. Comparisons between students in the balanced reading program to students in

---

[7] For example, higher level meaning of text, writing instruction and the presence of accountable talk, higher-level questioning about the meaning of text.

schools that had not implemented it yet and schools in various implementation stages suggest the effectiveness of a balanced reading approach for students who struggle to read. Descriptive analyses of posttest measurements reveal that 40% of the kindergarten students required supplemental services pre-implementation, but the percentage of students requiring these services decreased to 19% over the next three years. The first six schools that implemented the program demonstrated a statistically significantly lower need for supplemental services than the last six schools to begin their balanced reading program. Further, after only one year of implementation, supplemental services for all schools were considerably reduced.

Guthrie et al. (2001) find that classrooms with teachers providing more balanced reading instruction have higher achieving and engaged readers than students from classes with fewer opportunities for engaging in their reading interests. Results from a sample size of 576 grade 4 students lead Guthrie et al. (2001) with confidence to generalize their findings statewide. By controlling for mothers' levels of education, student gender, and amount of engaged reading (three variables that correlate significantly with achievement) and using between-group variables of balanced reading instruction and time spent reading, the researchers conclude that balanced reading instruction has a significant effect on student reading achievement.

Contrary to Bitter et al. (2009), Donat, (2005), and Guthrie et al. (2001), Diamond and Ongwuebuzie (2000) exploring the effect of a newly implemented balanced reading program on 2127 students enrolled in four k-3 schools and two 4th and 5th grade schools in the state of Georgia is not as supportive of the balanced reading approach. Diamond and Onguewubuzie (2000) find that balanced reading is associated with lower reading achievement and less favourable attitudes toward reading. They find that overall, students in grade 1 demonstrate the greatest reading achievement and that reading achievement generally decreases from grade 2 to grade 5 with a slight bump in achievement in grade 4 before dipping again in grade 5. Researchers caution that because this was a recent implementation, longer-term findings might be different.

What is lacking in this balanced reading research are comparisons with other approaches to early reading, as was done by the code-based and whole language research presented earlier in the review (Gee, 1995; Stahl & Miller, 1989, Stahl et al., 1994; Suggate, 2010, 2016). However, balanced reading instruction can be considered an approach to early reading that borrows from code-based and whole reading. Additionally, the literature review presented above indicates that balanced reading has a positive impact on many early reading students' achievement, particularly for students struggling to read. A balanced approach decreases the need for supplemental services for struggling students in one state, increases reading outcomes for all ranges of achieving students but conversely can be considered a strong reason for decreased test scores and enjoyment of reading for students in a school district from another state (Bitter et al., 2009; Diamond & Onguewubuzie, 2000; Donat, 2006). This review of the literature demonstrates that balanced approaches to reading contribute positively to the early reading outcomes of some students (see Bitter et al., 2009; Donat, 2006; Guthrie et al., 2001) but negatively influence others (Diamond & Ongwuebuzie, 2000).

Discussion

Further, as in the case of Maddox and Feng, students are assessed on limited elements of reading (spelling and reading fluency) and involve only one classroom, yet results of statistical significance ascribe a sense of generalizability and transferability. In terms of findings from the balanced reading research, these designs comprise quantitative analyses where the researchers admit their research designs omit certain elements inherent in teaching reading and student learning (Bitter et al., 2009; Diamond & Onwuegbuzie, 2000; Donat, 2006; Guthrie et al., 2001).

Examples include: the differential selection of teachers (see Diamond & Onwuegbuzie) where

increased numbers of teachers in experimental studies increase the likelihood of partial

implementation of instructional innovations; Donat finds that a comparison group of teachers

noticing the effects of a balanced reading program sought ways to learn about it, implement it

and in doing so weakened experimental-control comparisons; and Guthrie et al. and Bitter et al.

acknowledge that their data collection methods limit their understanding of the complexities

involved in instruction and student learning with Bitter et al. admitting their observation of

classroom instruction considers only a small portion of what might be happening in reading and

Guthrie et al. conceding their survey could not represent a clear picture of classroom practices

and reading activities. This critique of these researchers' designs is not an attempt to devalue

their findings but is an attempt to identify and address a gap in early reading research: the need

for emergent research designs concentrating on drawing out more holistic interpretations.

Whether teachers employ code-based, whole language or balanced approaches to reading

instruction, teachers must have the skills and knowledge to provide instruction that suits their

students' particular reading needs. The next section presents the state of preservice teachers',

their instructors', and inservice teachers' foundational early reading knowledge.

Teacher Knowledge

In broad terms, teacher change refers to increases in teacher knowledge in an area that

alters and improves professional practices and increases their students' achievement (Guskey,

2002; Timperley et al., 2007). Before moving towards the state of early reading professional

learning research, the ensuing section of the literature review explores how teachers'

assumptions influence their perceptions of struggling readers, what level of early reading knowledge teachers possess at preservice and inservice stages, and the role of experience and qualifications on teacher knowledge. Considering teachers' beliefs and teachers' overall knowledge in early reading is important as widespread teacher change is less likely if teachers do not know how their knowledge and assumptions influence their instruction (Borko & Putnam, 1995; Guskey, 1988, 2002).

Teacher Knowledge about Early Reading Instruction: Preservice Level

Perceptions of responsibility for struggling early readers. Though limited in quantity, it is important to present the extant research on the explicit and implicit perceptions evident within preservice programs towards the role of teachers in supporting students struggling with early reading. This is especially important as multi-tiered learner-centred models such as response to intervention are becoming more popular across Canada with expectations for classroom teachers to provide research-supported reading instruction and assessment differentiated to effectively meet the needs of all their students before specialized assistance outside of their classroom is offered (Government of Alberta, 2019; McIntosh et al., 2011).

One aspect of Joshi et al.'s (2009) research reports the perceptions 40 university instructors of early reading hold towards the factors they believe contribute to students' early reading difficulties. Not one of the instructors in their survey attributes the difficulties that struggling readers face to the quality of classroom instruction; instead, they cite the three most common factors: socioeconomic status, family background, and English as a second language background. Scharlach's (2008) research demonstrates that preservice teachers hold similar attitudes. In her four-month qualitative multiple case study exploring attitudes of preservice teachers, Scharlach's analysis uncovers the following themes regarding motivation, responsibility and self-efficacy to improve reading achievement for struggling readers: (a) reading is a developmental process aided by exposure to print; (b) in all probability students, who continue to struggle with reading have a biological disability; (c) parents need to be more involved; (d) at school, it is mainly the responsibility of the resource teacher to improve reading outcomes for struggling students. Findings from Joshi et al. and Scharlach are important to consider because if prevailing perceptions of university instructors (implicitly in the case of Joshi et al.) and preservice teachers (explicitly in the case of Scharlach) do not focus more upon the teachers' role in assisting struggling early readers, where will the impetus for developing one's knowledge to help these students take root? Thus if perceptions of responsibility are low, the knowledge that teachers will need to help struggling students may be lacking; findings which the following literature espouses.

Knowledge about early reading essentials. Joshi et al. (2009) and Washburn et al. (2011) suggest that many university preservice education programs do not align with current reading research. For instance, Joshi et al. find that many instructors lack understanding of core early reading principles and Washburn et al. show that university preservice education classes are not facilitating the core early reading skills their preservice teachers will require for providing effective early reading instruction. Their research indicates that many instructors, preservice teachers and inservice teachers perceive they know more than they do about early reading and that this likely contributes to the large number of students struggling to learn to read.

Washburn et al.'s (2011) research also highlights areas where preservice teachers lack knowledge. Ninety-one preservice teachers surveyed on their code-based knowledge and on their perceived ability to teach typical and struggling readers resulted in quite low scores.[8] On one

---

[8]Code-based knowledge assessed included phonological awareness, phonemic awareness, alphabetic principle/phonics, and morphology (Washburn et al., 2011).

32

knowledge test, the mean score on a phonics measure was 45%, with few participants having

explicit knowledge of terminology associated with phonics instruction and knowledge of phonics

principles for teaching decoding. Further, results from comparisons of perceived knowledge to

demonstrated knowledge and skill reveal that in every component except for phonics instruction,

preservice teachers believe their knowledge in these areas is higher than it is. Results from

research that moves out of the preservice stage and into the state of inservice teachers indicate

that this trend continues.

Teacher Knowledge about Early Reading Instruction: Inservice Level

Three studies illustrate discrepancies existing in teacher knowledge at the inservice level,

both in absolute terms and relative to preservice. Spear-Swerling et al. (2005), Mather et al.

(2001), and Cunningham et al. (2004) all find that teachers tend to be overconfident of their early

reading knowledge in comparison to actual measures about the skills that early reading research

espouses.

In their comparison of 131 experienced and 293 preservice teachers, Mather et al. (2001)

report that inservice teachers are more confident in their code-based instruction than preservice

teachers. However, on measures of reading knowledge essential for teaching struggling readers,

both groups of teachers demonstrate relatively low knowledge. Similarly, a few years later,

Cunningham et al. (2004) reveal the imbalance between teachers' perceptions of foundational

elements in early reading instruction and their actual knowledge. Measuring 722 kindergarten to

grade 3 teachers' knowledge of children's literature and code-based early reading skills and

comparing their knowledge to early reading measures of self-perception results indicate that 90%

of teachers are unfamiliar with many popular children's books and that 72% could not answer

more than half of the knowledge items correctly. Cunningham et al. find that teachers' self-perceptions of knowledge are overinflated with categorized high-perceivers scoring significantly lower on code-based tasks than the low-perceivers suggesting that even though many teachers believe they are prepared to aid struggling readers, for many this is likely not the case. One question not explored in Cunningham et al.'s work is how qualification and experience correlate with inservice teachers' perceived knowledge and measured knowledge about early reading.

Spear-Swerling et al.'s (2005) research indicates that teaching experience and level of qualification impact inservice teachers' depth of reading knowledge. After sorting 132 teachers into a high, medium or low group classified by varying qualifications, experience and training, they find that the high-background group consistently demonstrates higher scores on a perceived and actual knowledge test. In contrast, the low-background group score significantly lower than the other two groups. However, comparable to Mather et al. (2001) and Cunningham et al. (2004), the qualified and more experienced teachers, while outperforming less qualified and experienced teachers, score relatively low on questions on code-based early reading. This research on teacher knowledge is crucial because it highlights that many teachers, regardless of qualification and experience, require inservice code-based early reading professional learning. For the most part, this is where early reading professional learning programs focus.

Professional Learning Research

The literature on teacher change and its relation to professional learning has been a topic of discussion in educational research for three decades. Professional learning practices have evolved from one-off out of school conference-style workshops to include more school-based, situative, continuous and reflective professional learning communities (Fullan, 2007; Guskey, 1986; Hargreaves & O'Connor, 2018; Vescio et al., 2008; Webster-Wright, 2009). Producing long-lasting changes in teachers' practice is challenging due to the diverse experiences, attitudes,

34

and knowledge that individuals bring to their profession (Timperley et al., 2007). Prominent professional learning researcher Guskey found that systematic efforts are essential to change beliefs, attitudes and classroom practices (Guskey, 1988; 2002). Timperley et al.'s (2007) findings emerging from their best evidence synthesis section on teacher professional learning and development in literacy instruction indicate that though many teachers are committed to learning new theory and applying instructional practices, many continue to rely on method and theory not in alignment with research-supported evidence. Opfer and Pedder (2011), reviewing the general state of teacher learning and change through a complexivist lens, uncover contextually focused reasons for successful and unsuccessful professional learning: demonstrating how the elements of three systems they identified (the teacher, the school and the learning activities) interact and combine in different ways and with varying intensities to influence teacher learning. Though they illustrate the strengths in seeking to explain causality within the broader context of the school, most early reading professional learning literature encompasses linear cause and effect relations. The remainder of this section provides a review of the literature on quasi-experimental and comparative professional learning approaches and programs that provide teachers with instruction to develop their early reading knowledge at the pre and inservice levels.

Early Reading Professional Learning: Preservice Level

The literature discussing the role of professional learning for influencing the quality of instruction that early reading preservice teachers receive is minimal. However, one long-term study provides a perspective on how professional learning can improve instructors' and preservice teachers' early reading knowledge. Binks-Cantrell et al. (2012) studied the impact of long-term (three years) professional learning promoting empirically-based reading research on instructors' and their preservice teachers' knowledge of skills needed for effectively teaching

35

struggling students. Results from their survey measuring understanding of basic language constructs[9] demonstrate a significant difference between the experimental group (48 professors) and their preservice teachers) and the control group (66 instructors and their preservice teachers). The comparative analysis indicates that preservice teachers with professors in the experimental group scored significantly higher than their peers in all areas except one.[10] This research is relevant because it indicates that change can happen at a preservice level when professional learning sustains for a lengthy period and focuses on areas that the research indicates teachers need development in, such as learning about early reading instruction.

Early Reading Professional Learning: Inservice Level

Several studies explore the impact of professional learning approaches on teacher knowledge of early reading concepts and the effect of new learning on their students' early reading achievement. These studies demonstrate the importance of learning core early reading skills in expertly guided professional learning collaboratively and with mentors. The combination of learning core skills, expert leadership and collaborative learning effectively supports teachers' early reading knowledge, perceptions of their practice and improves student reading achievement (Brady et al., 2009; Carlisle et al., 2011; McCutchen et al., 2002; Podhajski et al., 2009).

Sustained professional learning coordinated by literacy experts, mentors, and research teams contribute to increases in teachers' knowledge and self-perceptions as reading teachers (Brady et al., 2009; Carlisle et al., 2011, McCutchen et al., 2002; Podhajski et al., 2009). Brady et al. 's (2009) demonstrated that teachers with mentors who provide feedback, model strategies

---

[9] Measures of phonology, phonics, and morphology.
[10] Preservice teachers did not differ significantly in a measure of phonological ability items.

and offer ongoing support correlate significantly with teachers' growth in early reading knowledge over the year, especially for teachers who initially scored low on pretest measures. Further, McCutchen et al. (2002) compare the reading knowledge of an experimental group of 23 teachers receiving two weeks of intensive training at the beginning of the school year along with visits throughout the year, to a control group of 20 teachers. McCutchen et al. find that with this sustained professional learning deepens teachers' knowledge of phonological awareness, teachers can use that knowledge to change their classroom practice and that changes in teacher knowledge and classroom practice can improve student learning. These findings are confirmed in a small study by Podhajski et al. (2009), focusing on first and second-grade teachers ($N = 4$) where coursework on explicit code-based early reading instruction had a significant impact on teacher knowledge and student outcomes. Teachers' engagement in their learning and teaching appears to increase when pairing early-reading professional learning with a sustained relationship with a reading expert. For example, though Carlisle et al. (2011) relate how all 111 teachers in their research developed their understanding and self-efficacy in teaching early reading, one-third of participants with literacy coaches rated their professional learning more useful and engaging than the two groups without one.

The four studies above demonstrate the instrumental role of mentors in helping teachers develop their reading instruction. The value of learning with peers across time, in conjunction with sustained professional learning, also appears as an essential ingredient for improving teachers' perceptions towards their learning and their knowledge of the early reading process (Gersten et al. 2010). Gersten et al. (2010) focus on reading comprehension and vocabulary instruction exploring the differences between 39 teachers in an experimental study group to 42 teachers in a control group. The control and experimental group took part in the same reading initiative and have similar professional learning, but the experimental group also receive sixteen 75-minute small group interactions held twice a month during the school year. Teacher knowledge, classroom practices and student reading achievement were measured. Findings reveal that the experimental group scored higher on comprehension and vocabulary knowledge measures relative to the control group with the effect on vocabulary instruction being statistically significant (d =.73). Appraisals from the experimental group on their professional learning show that 97% of these participants perceive that their professional learning was much more useful than their previous professional learning experiences.

Discussion

The findings outlined in this section of the review unpacks literature on teacher (preservice and inservice) knowledge, learning and change. This research indicates (implicitly and explicitly) that there exists a strong need for professional learning that can create long-lasting changes to teachers' early reading knowledge that improves their instruction and subsequently help students who are struggling to read. Many teachers at the preservice level demonstrate a lack of code-based knowledge of early reading principles (Joshi et al., 2009; Washburn et al., 2011). Research on inservice teachers' knowledge indicates that though teacher qualification and experience correlates with increases in student early reading achievement (see Spear-Swerling et al., 2005) many early reading teachers do not possess a vast wealth of early reading knowledge (Cunningham et al., 2004; Mather et al., 2001; Spear-Swerling et al., 2005). However designing learning opportunities for instructors in preservice education programs and inservice teachers can develop their early reading knowledge and subsequently improve early reading outcomes for primary grade students (Brady et al., 2009; Carlisle et al., 2011; Gersten et al., 2010; McCutchen et al., 2002; Podhajksi et al., 2009).

There are several gaps and limits in the early reading teacher knowledge and learning literature. The majority of the research in the field of early reading (instructional approaches, teacher knowledge, and teacher change), is quantitative and focuses on documenting the lack of code-based knowledge teachers have and on preordinate structures and designs of professional learning and reading interventions. This research communicates the impact of quasi- and experimental research, yet understanding what teachers are realistically undertaking in their professional learning and the opportunities presented to them is underexplored. The majority of these research designs synthesized in the review of the studies above omit deep considerations of the complex variables interacting in school systems and, therefore, silence how contextual factors influence teachers' early reading professional learning and teachers' knowledge and classroom instruction. This current study explores teachers' professional learning through a complexivist lens to seek a more holistic interpretation and understanding of context.

Complexity Theory as a Conceptual Framework

> Complexity thinking might be positioned somewhere between a belief in a fixed and fully knowable universe and a fear the meaning and reality are so dynamic that attempts to explicate are little more than self-delusions. In fact, complexity thinking commits to neither of these extremes, but listens to both. (Davis & Sumara, 2008, p. 35)

Self-Organization and Emergence, Nestedness, Complexity Reduction

Although this section separates and describes complexity theory through these terms below, self-organization, emergence, nestedness and complexity reduction are theoretically and practically interacting components. Morrison (2008) outlines that a central way to consider complex systems is their characteristics of self-organization in which changes in systems emerge from within and outwards and in reactions to the varying strengths of tensions across other systems (termed autocatalysis). For Morrison, the central features of self-organizing systems are their adaptability, openness, ability to learn from and react to feedback through inward and outward communication, the ongoing emergence through change and interaction in disequilibrium and the infinite connectedness that varies to proximity, and their concern and relatability with and to other systems. In this way, education is a system comprised of and nested within and across all social systems.

This research applies three main elements of complexity theory for considering the interactions at play in the educational system: (1a) self-organization and (1b) emergence, which Morrison (2008) considers partners to each other. For a system to self-organize infers that change is happening, and the change that happens is the newly emergent system. Emergence then creates a force through the momentums that have changed it and this change (or emergence) moves in different directions that informs and influences itself and other systems that are intimately connected to it; and those systems to other systems in varying degrees and in continuum. This relates to another feature of complexity theory, the notion of nestedness; (2) the section on nestedness illuminates the importance of understanding the place of early reading within varying systems, which connects to the system that early reading teachers, their students, the school and the board are in. Bringing to the surface here the nature of nestedness ensures that a consideration of relationships between systems are implicated in the search to unpack the influence of context upon varying systems that are interconnected but also likely have differing self-interests interacting within and upon each other; (3) understanding complexity reduction is central to guiding this research project to illuminate how contextual variables interact and influence teachers' learning. Research with complexity reduction explicitly in the forefront helps bring back purposes of education and balance what is, with what could be by ensuring context is

40

opened up, considered and valued so that self-organization and emergence reflect a need being met rather than a reaction to presumed norms in which teachers and students work, live, and learn.

Self-organization and emergence. According to Morrison (2008), self-organization is a central pillar of complexity theory and that "local circumstances dictate the nature of the emerging self-organization" (p. 20). Each system possesses its uniqueness and identity, and these characteristics enable systems to renew and perpetuate. A principle of self-organization is that for emergence to be possible systems require a transcendental quality that naturally belies the ability to change and influence change (Morrison, 2008; Koopmans, 2017). Each system that is a part of another system has a homeostatic intent and systems nested within other systems where non-linearity prevails challenges conclusions of specific cause and effect. Koopmans (2017) suggests shifting the aims of research from a static quest for momentary results at a specific time to a descriptive process of research and analysis seeking to understand the process of transformation through how systems process their self-organization. Nestedness is discussed below, though it is important to point out here that within systems there are many other related systems. In my research, I am looking at teachers who are nested in their classrooms, nested in their schools, and nested in a school board that is nested in a province. Along with these blunt categorizations, there are many other systems that individuals and groups are nested within.

An aim of this research is to understand how professional learning is perceived to be influencing early reading instruction for teachers in three different schools that comprise students coming from distinctly different communities and therefore having different background knowledge and experience. Each class, school, and the board itself are different but connected systems. Therefore these varying open systems who possess unique characteristics are also involved in similar professional learning opportunities that will vary in how well they meet their students' early reading needs. From this the emergence of knowledge and practice of reading is going to be perceived in varying ways with varying degrees of effective practices emerging. Morrison (2008) describes the adapting and renewing natures of systems are what preserves systems over time. However, by considering how differently situated systems (e.g., different schools in a board) emerge and self-organize based on the availability and accessibility of early reading professional learning, it is likely that the status quo will have some schools trending in a positive direction. However, for others the opportunities are restricting opportunities to develop in areas of need.

Research that considers the self-organization of contextually different systems (e.g., pairs of teachers in demographically different schools) can illuminate how learning and practice are perceived to have emerged from past experiences. Then, research can bring the local perspectives together so that larger systems can reflect the needs of systems that are similar (grades, subjects) but also inherently different (because of context and influence of demographics). Opening up systems to better understand the role of context on self-organization and emergence allows (or at least eases towards) more effective learning experiences for teachers and students to emerge (Davis et al., 2012).

Nestedness. Bridging complexity theory with critical realism, Cochran-Smith et al. (2014) acknowledge the strength in using complexity theory for contributing to "a unique platform for teacher education research, which has theoretical consistency, methodological integrity, and practical significance" (p. 2). Rather than using a traditional linear, cause and effect investigation, Cochran-Smith et al. (2014) recognize complexity theory's versatility for holistically exploring and interpreting the varying effects of the complex interactions within systems, nested within systems. The notion of nestedness denotes that entities exist in systems, yet also give rise to systems (Davis & Simmt, 2006; Davis & Sumara, 2008, 2012). For example, one group of teachers' professional learning needs are given priority over others and from this certain practices are improved while others are neglected with these actions influencing the development of different aspects of students' learning and teacher change.

Davis and Simmt (2006) offer a theoretical discussion of complexity science as a framework for interpreting teachers' mathematics professional learning. They present the notion of four interacting aspects of their learning system and how it opens up new possibilities for considering new learning experiences for preservice teachers. Labelling four aspects (mathematical objects, curriculum structures, classroom collectivity, and subjective understanding), they consider how each aspect is nested in one another. In a similar notion, my research is framed to conceptualize the contextual variables that relate to teacher early reading learning by considering how nestedness enables and constrains their learning opportunities (in their school, in their board, in relation to responsibilities that are parts of their other systems). Davis and Simmt relate the notion of nestedness and the boundaries between nested systems:

> complex forms are often nested, with many intermediate layers of organization, any of
> which might be properly identified as complex and all of which influence (both enabling
> and constraining) one another. Complexity science prompts attentions toward several
> dynamic, co-implicated, and integrated levels – including the neurological, the
> experiential, the contextual/material, the social, the symbolic, the cultural, and the
> ecological – rather than isolated phenomena. (p. 296).

Though Davis and Simmt are relating how the nested and interconnected nature of inservice teaching and learning relate to learning about mathematics concepts and instructional practice, it

43

illuminates the need to be considerate of the different levels of nestedness of social learning

systems within one domain of curricular knowledge.

In my research I am interested in the perceptions of experiences of teachers who teach

early-reading (and also teach a number of other subjects) and are nested in classrooms with their

particularities, and schools with their particularities, and boards with their particularities but also

are nested in their social systems (e.g., families and other communities that have their different

particularities). I am interested in unpacking how these interacting systems influence instruction

and how the system that these teachers work and live in might be better considered so that the

learning needs of their students who are nested in their own particularities can be more

empathetically met. Therefore, a focus on nested systems is an element of complexity theory

made explicit here and is at the forefront of my complexivist conceptual framework and

research.

Cochran-Smith et al. (2014) discussing social learning systems such as teacher education

emphasizes the importance of ensuring that social systems should not be considered

hierarchically; that social systems are parts of sets of social systems that are nested within and

amongst each other and therefore have to be considered holistically rather than individually or as

top-down in order of presumed importance. It could be considered that the current system is

hierarchical and has existed in educational research and policy for too long (see Gough, 2012)

and that a complexivist led educational research approach re-conceptualizes and attempts to

balance the hierarchical experiences that many in the system confront. Power imbalance and

traditional structures are discussed next in terms of complexity reduction.

Complexity reduction. Some aspects of interactions within systems are considered

impediments or controlling agents over possibilities for what could be argued as more natural

44

emergences (Fenwick, 2012; Hetherington, 2013). This unnatural process is considered

complexity reduction and is an unavoidable aspect of social semiotic systems where the push and

pull (or exchanges) that happen occur between levels in the realm of meaning making (Biesta,

2010; Hetherington, 2013). Complexity reduction refers to actions from elements within systems

that impact how emergence is directed (Hetherington, 2013). By limiting options for actions

between elements in a system or through measures used to control language within a system

emerging possibilities are restricted (Hetherington, 2013). Aspects of school systems such as

"timetables, curricula, classroom organization and layout, and school hierarchies may all

contribute to the reduction of the potential complexity of schooling, or mitigate against the

conditions for emergence in various ways" (Hetherington, 2013, p. 74).

Considering the nature and issues of and resulting from complexity reduction, a

complexity theory approach can work towards uncovering half (or at least partial)

understandings (Davis & Sumara, 2012). Making explicit the availability of options that teachers

have in their professional learning provides a way to show that what might be assumed to work

for the whole, is in fact, only effective for a few (Biesta, 2010). This is a central reason for

utilizing the notion of complexity theory in educational research and ensures that the

perspectives of those responsible for ensuring students are having their needs met in contextually

considerate ways. Theoretically and practically, this provides a path for educational research to

open up some of the current limitations teachers appear to be facing in relation to their learning

opportunities and how this, in turn, influences their instruction and the accessibility to the

learning environment that their students need (Biesta, 2010). Over time, placement in systems,

such as schools that are nested in systems that are limited by provincial oversight bodies, a

certain kind of learning opportunity for everyone that is part of the school system is shaped and over time considered or assumed to be the norm. In this environment, a certain type of top-down normalcy is assumed and what is reducing equitable opportunities for students and teachers is less visible and is not normal:

> Rather than to think of such regularities as naturally occurring phenomena, they are actually in the most literal sense social constructions. To say that these are social constructions is neither to say they are good nor they are bad. While in some places complexity reduction can be beneficial, in other cases it can be restraining. But since any attempt to reduce the number of available options for action for the 'elements' within a system is about the exertion of power, complexity reduction should therefore be understood as a political act. (Biesta, 2010, p. 498)

Biesta continues along these lines as he questions the purpose of evidence-based educational practices. Before limiting the nature of teaching and learning by using prescribed practices – or in this research's goals, teachers' learning opportunities which in turn affects their students' learning – the educationally-connected body/ies (teachers, students, administrators, parents, governments, researchers) need/s to when considering the teleological character of education make their/our first question to be questions of purpose. What is the purpose of doing what we are doing, and how is this equitable for our learners, or how is it not?

Therefore in my case, where an attempt to understand why too many students are continuing to fail to learn how to read proficiently, considering what realm of factors reduces complexity reduction provides balanced considerations of these tensions and possibilities. The role of context and how limitations that teachers face depending on the degree that theirs and their students' contextual uniquenesses are not being considered through the professional learning opportunities that are available (or are not) is opened up by showing how teachers who are trying to meet the needs of students in urban or rural lower socioeconomic communities than say a suburban middle-class community likely exhibit quite contrasting but also comparable needs.

Discussion: Complexity Theory and Educational Research

This doctoral research study is conceptualized through a complexity theory lens. Complexity thinking underpins the literature review, the research questions and the methodology. Epistemologically, complexity theory positions research and researcher as embracive and inclusive of the uniqueness that emerges in the collected perspectives of those most directly influenced within a particular context and topic. As a metatheory, complexity theory represents a paradigm shift in how to consider and conceptualize systems more holistically; and a way forward in researching "discontinuities, ruptures and emergence instead of relying on commonalities and the central tendency in groups" (Koopmans, 2014, p. 26). Complex systems can be understood as self-organized, composed of dynamic elements transmitting and receiving varying strengths of interactions working together and in opposition, from which new forms of action or information emerge as unique to that specific system (Morrison, 2008). Focused on the interactions between the components comprising the system, complexity theory is well situated to be used as a dynamic form for conceptualizing and developing a multifaceted knowledge about the elements connected to the context of professional learning and early reading instruction (Cochran-Smith et al., 2014). Systems, as considered through the lens of complexity theory are self-organizing bottom-up processes (Mason, 2008; Morrison, 2008). Citing the unpredictable nature of systems, a process of exploring systems from a bottom-up perspective reveals more profound ways for representing the relatedness of interactions and emergence (Osberg et al., 2008).

Complexity theorists acknowledge the dynamic role context plays in uniquely shaping each system (see Byrne, 2001; Davis & Sumara, 2012; Haggis, 2008; Kuhn, 2008). Within this project's complexity theory lens, context is considered as a combination of initial conditions, with history through time acting on and in current particular conditions. This tension gives rise to ever-adapting systems shaped in their "specific history of emergent and ongoing conditions" (Haggis, 2009, p.54). Complexity theorists accept that all systems are unique, connected and ever-changing (Osberg et al., 2008). Particular configurations of the present system are also connected with other systems, influenced and influencing, so that the effects that emerge, though connected with other systems, are also always importantly unique (Haggis, 2009). Considering systems as emergent entities sheds light on how emergent needs within systems are reflected by understanding a system's proximity to varying contextual environments.

In relation to a school system, this speaks to how children are linked to families, teachers, peers, society (local and at-large) and groups with varying strengths of interactions on and amongst all agents and elements (Morrison, 2008). Looking at this in another, but still connected level, teachers are linked to other teachers, support agencies, policy-making bodies, funding bodies, and governing bodies (Cohen et al., 2011). Considering that each individual that is connected to this system and is also connected with many other elements that are parts of other systems it is reasonable to expect to find unique aspects in different schools guided in large part by particular, local contexts (Osberg et al., 2008).

A strength in using complexity theory is how it embraces the dynamic and emergent natures of and in systems, including teachers as co-learners and co-constructors of knowledge (Morrison, 2008). An inherent responsibility that most elementary teachers have is providing instruction across a spectrum of topics. It is understood that teachers will vary in their experiences, skills and knowledge compared with colleagues (Gardner, 2006). To consider, in one school, teachers will differ in years of experience, styles and approaches to teaching, and in many other ways that explicitly and implicitly shape and influence theirs and, in some aspects, their colleagues' instruction. School principals will have varying interpretations of what is most instructionally important for their teachers to focus on, which professional learning experiences are necessary, or how school personnel can be best utilized. This conjecture on the complex natures that co-exist within a system very likely captures a fraction of the complexities comprising a school system. Even in just one school, interpretation of what professional learning means, how it should be used, or the type of instructional approaches suited for different elements of early-reading instruction is likely to have elements that are understood similarly and differently across and between every educator (Guskey, 2002).

Moving forward. A complexity theory mindset implies positioning one's self from an epistemological standpoint in which knowledge is transactional, unfixed and "(becoming educated) is no longer about understanding a finished universe, or even about participating in a finished and stable universe. It is the result, rather, of participating in the creation of an unfinished universe" (Osberg et al., 2008, p. 215). A research design that deeply interprets the processes of interactions appearing to underpin early reading teachers' professional learning can clarify aspects of uniqueness (similarity and difference). Understanding how context is at play and related to specific groups of teachers in multiple schools in the same board offers a way to consider the richness of the features emerging out of different self-organizing systems. How does physical distance from the central school board office and teaching a student population whose first language is not the first language of instruction influence teachers' perspectives towards their learning experiences and opportunities? How does teaching in an urban core influence students' and teachers' early reading needs? How does the environment appear to influence learning needs when you are not isolated or urban? What kinds of patterns in perspectives are there across schools, and how can professional learning address similarity and difference? Approaching these kinds of questions from a complexity theory lens opens up the nature of self-organization of systems nested in systems, represents visible and seemingly invisible forms of complexity reduction and how learning and practice are influenced, and provides a textured snapshot of the emergent and emerging interactions between variables influencing the teaching and learning process. Challenges about a complexity theory-informed research are raised by Cochran-Smith et al. (2014) because this kind of research is descriptive and rejects causal implications; they suggest that complexity theory can be used as the framework for qualitative educational research that can help to empirically trace the learning processes teachers interact within and use this information to enhance the learning of teachers and all students. In the next chapter, I describe the methodological underpinnings of my research.

Chapter 3 – Methodology

Schools have unique attributes, the result of daily interactions amongst staff and students, with each person having a different role in the collective system constituting the school. Investigating how teachers perceive their learning opportunities, how complex elements within their schools impact the types of learning experiences they have, and whether and how these influence their reading instruction fits well with a complexity theory qualitative case study approach (Cohen et al., 2011; Hetherington, 2013). A more inclusive portrayal of the state of early reading instruction and teacher learning is generated from framing research in this fashion and from this a new stance for viewing teachers' professional learning. This qualitative complexity theory framework for portraying the non-linear and dynamic perceptions teachers have of their professional learning experiences and how this influences their early reading instruction embraces the complex natures inherent in schools and school boards. Informed through the body of literature on the approaches to early reading, teacher knowledge and change, and early reading professional learning and guided in a complexity theory conceptual framework, two overarching research questions guide my research. I present these in the next section. I follow this up with a description of the details of the emergent qualitative research design I employ to answer the research questions.

Research Questions

To open up the exploration into the complexities of early reading instruction and learning, two broad research questions guide my qualitative research study:

> 1. How do contextual variables at the school, board, and provincial level influence the planning, delivery and uptake of early reading professional learning opportunities?

2. How do teachers perceive the relationships between (a) their professional

learning experiences, (b) their classroom early reading practices, and (c) student

reading outcomes?

Qualitative Research Methodology

An epistemological assumption inherent in qualitative research is its emerging approach

to inquiry (Denzin & Lincoln, 2005). Creswell (2007) remarks how collecting data in a natural

setting is sensitive to the local context, and that data analysis that works towards establishing

patterns and themes is best suited to reflect the voices of participants by providing a complex

description and interpretation of the phenomenon under study. The roles of educators in schools

across North America are generally the same (e.g., principals, teachers, students), yet individual

schools are going to differ in many aspects because of the demographics of their particular

environment. Therefore each school will retain its uniqueness. Focused on how people construct

meaning from their experiences, qualitative research seeks deep, rich meaning from interpreting

people's experiences in their specific, natural settings (Merriam, 2009).

The choice of a qualitative case study research design is due to its capacity to open up

depth to a topic at localized levels for drawing out meaningful abstractions across multiple

settings (Merriam, 2009; Seidman, 2019). In my research, I seek to understand the complexity of

the variables interacting upon and within teachers' professional learning, early reading practices

and students' reading achievement. The purpose of my research is not to evaluate what specific

professional learning improves practice and increases student reading achievement. Rather, my

research design contributes to reading research and teacher learning by conceptualizing how

context influences learning opportunities and teaching practice. Through my qualitative research design, I seek to understand how contextually conceptualizing teachers' early reading professional learning can lead to real change and reductions of reading failure.

Case Study Design

The research design follows an instrumental multiple case study structure. Stake (1995) differentiates between intrinsic and instrumental cases. Intrinsic case studies have a particular case as an object of interest. Instrumental case study refers to a way where seeking to understand a particular phenomenon is above and beyond the case itself (Grandy, 2010; Stake, 1995). My research intends to understand how teachers' early reading professional learning is perceived to influence their instruction and their students' reading outcomes. Underlying this phenomenon is an interest in understanding how variables are interacting to influence teacher learning and practice. The design is characterized as multiple because I seek to understand the phenomenon beyond one bounded case (Chmiliar, 2010). This multiple instrumental case study extends the depth of understanding of the phenomenon of teachers' professional learning through exploring early reading teachers within multiple schools and works to draw out the uniquenesses and abstractions from three different schools across the same school board.

This instrumental qualitative case study design comprises two teachers nested within three schools. It involves an exploration of interview data from teachers, school principals and school board early reading experts: All participants in the position to have a deep sense of the process of early reading or the nature of professional learning within their particular schools and across the school board. Teachers are the central participants in my study. They are necessary for gathering perspectives on how their learning experiences influence their reading instruction, what elements and interactions influence their professional learning (school, board and beyond) and if and how they perceive these experiences impact their early reading instruction. The perceptions of school principals provide insight into the context of their schools, and what kinds of interactions appear to be at play and how these factors contribute to teachers' perceptions of their early reading knowledge and practice. School leadership forcefully shapes the school learning environment, and having an awareness of the leadership levels within schools is, therefore, essential to elaborate upon (Sider et al., 2017). The board personnel's perceptions of how the school board provides opportunities for early reading professional learning adds depth and provides contextual information supplementing the data from the teachers and principals at their particular schools.

Participant Recruitment

Purposeful sampling underlies the selection of early reading teachers, principals, board reading experts and the case schools. The "logic of purposeful sampling lies in selecting information-rich cases, with the objective of yielding insight and understanding of the phenomenon under investigation … in contrast to the random sampling procedures that characterize quantitative research" (Bloomberg & Volpe, 2008, p. 69). Further, the choice for purposeful sampling is anchored in striving for participants' perspectives reflective of their diverse teaching contexts and the reflective influences these have upon their perceptions of professional learning and practice (see Emmel, 2013, on maximum variation).

Gaining access to the board and school participants required following multiple steps; broadly, school board and school willingness and university ethics approval. Approximately two months following ethics application submission, I received official university approval to formally contact the school board and prospective research participants (see Appendix A, B, and C to view the university and school board ethics approval). A doctoral colleague with ties to the school board formally introduced me to the Director of Education. I sent a letter to the Director

54

of Education, cc'ing my colleague, introducing myself and my research intentions and needs. I asked for a sample of three schools differing in terms of geographic location, the linguistic context of the community, and socioeconomic status because purposeful sampling was a crucial part of my research design for unpacking contextual influences on teachers' learning. I also requested an interview with a board reading specialist for authentically understanding the perceptions of literacy and teacher learning within this level of the educational system. The Director of Education forwarded my request to the Director of Human Resources (DHR) who was the gatekeeper to conducting research in the board. The DHR responded that they would discuss my project with a group of principals to gauge their interest in participating. One week later, the DHR notified me that two schools were interested (Blending Elementary School and Outer Elementary School, one rural and one suburban). The principals were cc'd, followed up with, and research relationships were established (see Appendix D and E for the letters of recruitment letters that I shared with the principals and teachers).

With research permission granted, I also reached out to two early reading consultants who became participants SB1 and SB2 (see Appendix F for the school board recruitment letter). In the course of introductory emails, SB1 suggested that Central Elementary School was ideal for this research because of its particular urban demographic and its contrast to the two other participating schools. Formal introductions made to principals regarding the nature and needs of my research occurred first. The principals provided this information to their early reading teachers (see Appendix G, H and I for these consent forms). I received interest and agreement from two early reading teachers at each school.

SB3, the third school board participant, was added to my research after multiple participants suggested I interview her because of her depth of early reading knowledge and time spent working in the school board. Collectively, each participant in this study was provided ethics information, permission slips, reasons for the research, their particular role in the research, and guarantees of their rights as participants and assurance of confidentiality. Based on principles of purposeful sampling, and through the positive receptivity of those I contacted at the board, and school level, an important piece of the research design was in place.

Overview of Participants

An overview of the participants and their position within the board is presented in Table 2. The three school board participants are identified as SB1, SB2, and SB3. The six teachers and 3 principals are identified first by their school, second by their participant role, and third by their identifier as 1 or 2 for teachers. For example, the principal from Outer Elementary School is OP. The two teachers at OES are OT1 and OT2. Such identifiers maintain the awareness of the context participants speak from and shift attention from how randomly chosen pseudonyms might influence readers' initial perceptions and possible biases. Two of the three principal participants, all six teacher participants and the three board early reading participants are female. A scan of the elementary school websites from the case school board indicates that of the 213 kindergarten to grade 7 teachers, 85% are female. Specifically, 91% of grade 1 and 2 teachers are female, and in regards to principals, 64% are female. Data from the Government of Québec (2003) and Statistics Canada (2017) indicate that approximately 85% of elementary teachers are female. These descriptive statistics suggest that female participants' representation in this study aligns with female and male representations in the case school board and teachers, provincially and nationally.

Relevant contextual, demographic information about the nine participants from the three schools is in Table 2 (below), moving from board, then through schools. In-depth teacher profiles, voiced in the first-person, are in Appendixes J-O. These narratives provide a story of each teacher participant and offer the reader an intimate portrayal of their perceptions to and beyond this study's exploration. Below are truncated descriptions of each of the study's participants.

Table 2

Overview of Study Participants

Pseudonym	Position	Gender	Years in Current Role	Years of Teaching Experience	Years in Board	Additional Educational
Board Literacy Experts and Consultants						
SB1	Board Literacy Specialist	F	3	14	14	No
SB2	Low SES Schools Coordinator	F	1	13	25+	Reading Recovery
SB3	Literacy Consultant	F	8	30+	40	Reading Recovery
Blending Elementary School						
BT1	Gr 1 & 2 Teacher	F	10	20	15	No
BT2	Gr 1 Teacher	F	6	18	17	No
BP	Principal	F	3	5-6	22	Masters of Education
Outer Elementary School						
OT1	Gr 1 Teacher	F	4	5	4	3 Additional Qualifications (ESL, SPED, Early Reading)
OT2	Gr 2 Teacher	F	1	3	3	No
OP	Principal	F	6	15	+10	Unknown
Central Elementary School						
CT1	Gr. 2 Teacher	F	1	18	18	No

Pseudonym	Position	Gender	Years in Current Role	Years of Teaching Experience	Years in Board	Additional Educational
CT2	Gr. 1 Teacher	F	6	11	11	No
CP	Principal	M	7	7	21	Masters of Education

School Board Participants

SB1. At the time of our interviews, SB1 a third year full-time board literacy specialist worked out of the school board office to support the English Language Arts for all schools in the school board. Prior to this position, she was a classroom (grades 2, 4 and 6) and resource (grades 3-6) teacher for eleven years, as well as a coordinator of the board's new teacher induction program.

SB2. SB2's 31-year career spans 13 years as a teacher and 18 as principal. At university, she specialized in early childhood education. At the time of our interview, she was coordinator in this board for a provincial grant that allots funds towards helping teachers working in rural, low socioeconomic schools such as OES. She has been with the board since the early 1990s. SB2 is a trained Reading Recovery teacher, with the bulk of her teaching in kindergarten and grade one, though she has taught every elementary grade.

SB3. SB3 graduated in the 1970s with a bachelor of education and reading specialist designation. At the time of our interview, she was in her 40th year with the board. Most of her teaching experiences are in grade 2, 3 and 4 classrooms, and as a resource teacher, she provided Reading Recovery to students up to grade 6. At the time of our interview, she was retired from teaching, and was working with the board as a special education consultant and working with

SB1 to support teachers' implementation of two board-supported reading initiatives with most elementary schools across the board.

Blending Elementary School Participants

BT1: Teacher, Grade 1&2. Over her approximately 20-year teaching career, BT1 has taught pre-kindergarten to grade 4. Beginning her career in this case school board, she moved away to a neighbouring province to teach in a large city for several years before she and her family moved back. BT1 has had three children during her teaching career. Her teaching experiences consist of mostly grade 1 and 2, and her career goal is to work as a resource teacher.

BT2: Teacher, Grade 1. BT2 obtained her Bachelor of Education degree in the early 1990s. Following a short teaching contract and around the time of massive layoffs in her first school board in a neighbouring province, she decided to move to this board and school to teach a grade one-two class. BT2 taught kindergarten to grade six for just over ten years before going on an extended maternity leave from which she returned six years ago. She has taught in a 70% teaching capacity within the grade one language arts environment ever since.

BP: Principal. BP started with the board in 1995, teaching pre-kindergarten up to grade 3. Returning from maternity leave in the early 2000s, she took on a job comprising 80% teaching and 20% as administrator for four years before taking another maternity leave. Upon her return, she became a full-time principal at a middle and high school for eight years. At the time of our interview, BP had been the principal at BES for three years.

Outer Elementary School Participants

OT1: Teacher, Grade 1. Following her one-year teacher education program, OT1 moved to teach in a rural school in a western Canadian province where she taught for one year. After receiving news of the school's financial troubles and with no immediate prospects in that area, she returned home to an eastern Canadian city and completed three additional qualifications (ESL, Literacy and Special Education) before being hired at OES. At the time of our interview, OT1 was in her fourth year teaching at OES in the primary grades.

OT2: Teacher, Grade 2. OT2 began her career in the school board as a kindergarten teacher in a small rural school, spending the better part of two years there. At the time of our interview, OT2 was in her first year teaching grade 2 at OES.

OP: Principal. OP graduated from her bachelor of education program in the early 1990s. Her first teaching position was in her childhood elementary school, and having taught a decade and a half in primarily rural schools, she is currently in her sixth year as the principal of OES.

Central Elementary School Participants

CT1: Teacher, Grade 2. CT1 is in her eighteenth year as a teacher. She transferred from a grade 11 position in the rural school where she started her career to CES seventeen years ago, exclusively teaching grades 1-3. At the time of our interview, she was teaching grade 2. Across her teaching career, she has taken two maternity leaves.

CT2: Teacher, Grade 1. After completing a psychology degree, CT2 took a one year bachelor of education program and began teaching at CES in the mid-2000s. In her eleventh year at CES, she taught grades 2-6 in her first few years before being moved into a grade 1 classroom where she has taught the past six years.

CP: Principal. CP has been with the school board for 21 years: Fourteen years were spent teaching (7) and as a vice-principal (7) in the same high school. He has spent the past seven years as the principal of CES.

Data Collection: Interview Process

The choice to conduct in-depth interviews for the method of data collection is because of

the intimacy the structure provides participants for deeply sharing their perspectives on a

particular phenomenon of interest. Interviewing says Stake (1995), "is the main road to multiple

realities" (p. 64), able to access what cannot be observed. Connecting to social and educational

issues is effectively approached through interviewing (Seidman, 2006). The interview's value is

in how it draws out interviewees' perceptions: a synthesis of abstractions based on interviewees'

perceptions representing their interpretations of their educational and social experiences

(Seidman, 2006). Though other forms of data collection provide a way to approach the

experience under exploration, interviewing done well unpacks the meanings people construct

from their experiences (Seidman, 2006). Teacher and principal interviews took place in the

schools, usually in a resource or staff room, sometimes in a teacher's classroom, once in the

principal's office, and once in my home office. Each interview with board-level participants

occurred at the central board office. All teacher, principal and board interviews were completed

within five months. The structures of the interview process with participants are explained

below.

Teacher In-Depth Three-Series Interviews

The interview approach with teacher participants follows Seidman's (2006, 2013, 2019)

in-depth three-series design. This process comprises a stage-like approach to interviewing that

works to access teachers' lived experiences of early reading and early reading professional

learning. Each of the three interviews ranges from 60 to 90 minutes.

Interview 1: Focused life history. Questions in the first interview focus on participants'

reconstruction of their past learning experiences up to their present. Reconstruction occurs

through questions guided in a range of categories and connects to constructing a sense of participants' perspective on reading (learning and teaching) across their lives (Seidman, 2006).

Questions relate to teachers' past lives and up to their most recent teaching and learning

experiences. Delving into the past in the first interview provides a platform for understanding

how teachers interact and react to their past learning experiences. Questions to nudge

perspectives around reading and learning uncovered in the review of the literature (e.g.

motivation towards struggling readers, preparation to teach reading, the influence of professional

learning) illuminate how teachers perceive their experiences shape them and influence their

professional learning opportunities (see samples of open-ended questions and probes used in

Interview 1 in Appendix P).

Interview 2: The details of experience. The second interview opens up the "concrete

details of the participants' lived experiences" (Seidman, 2013, p. 21) within early reading

instruction and professional learning. The purpose of this interview is to have participants

provide details rather than opinions on their teaching and learning experiences. In this interview,

the compilation of a range of specific details informs the direction and focus for the third

interview in this process (see Appendix Q for probes used in Interview 2).

Interview 3: Reflection on the meaning. The third and final interview has participants

"reflect on the meaning of their experience ... [to address] the intellectual and emotional

connections between the participants' work and life" (Seidman, 2013, p. 22). This interview is

akin to a synthesis of the first two. It facilitates an experience with the participant to have the

participants consider how different factors are interacting and influence their early reading

teaching and professional learning: See Appendix R for the kinds of probes used in this

interview.

One hour focused semi-structured interviews with the three school principals, and board participants tease out how aspects of context appear to influence professional learning opportunities, early reading instruction and student achievement. The contextual data these participants provide contribute to illuminating the complexities interacting within the learning network at the board and in each school. Themes guiding interview probes for principals centre on their educational background, leadership characteristics, perspectives on school guidelines and expectations for professional learning, and reading instruction in general. The board participants responded to a list of questions regarding their knowledge of, experience with and perspectives on early reading instruction and professional learning (see Appendix S and T for the respective interview forums).

Data Analysis

An emergent and iterative process guides the data analysis (see Figure 2 below). According to Attinasi, Jr. (1991), a phenomenological inductive process is a place for concept and hypotheses to emerge out of the "the informants' lives reported in the interviews [with the] induction process … constrained only by the research perspective" (p. 3). In embracing the messiness of complexivist research methodology, the analysis occurs across eight stages involving multiple deep steps of exploration and continuous data reduction (e.g., Koopmans, 2014). Seidman's in-depth three-series interview process structures the deep probe into the lived experiences of participants. Following the suggestion of grounded theorists Glaser and Strauss (1999), I initiated a process of memoing in each stage of analysis to streamline the depth of data into their working categories. Initially, I transcribed the 24 interviews verbatim using Microsoft Word 2010. Transcripts were transferred to QSR International's NVivo 11 software for preliminary analysis and conceptual sorting before transferring the data back to MS Word for

most of the heavy work in the data analysis. Seidman's (2013) processes for data analysis of my

in-depth three-series interviews with teachers broadly structure my data analysis: (a) studying,

reducing and analyzing the text, (b) creating participant profiles and themes, and (c) making and

analyzing thematic connections. Narratives in the form of extensive profiles for teachers and

shorter vignettes for principals are the initial reductions of the data. They occur inductively

rather than deductively by allowing the first stage of analysis to provide emerging conceptual

categories (see Appendix J-O for the full profiles of the teacher participants).

Figure 2

Stages of Data Analysis

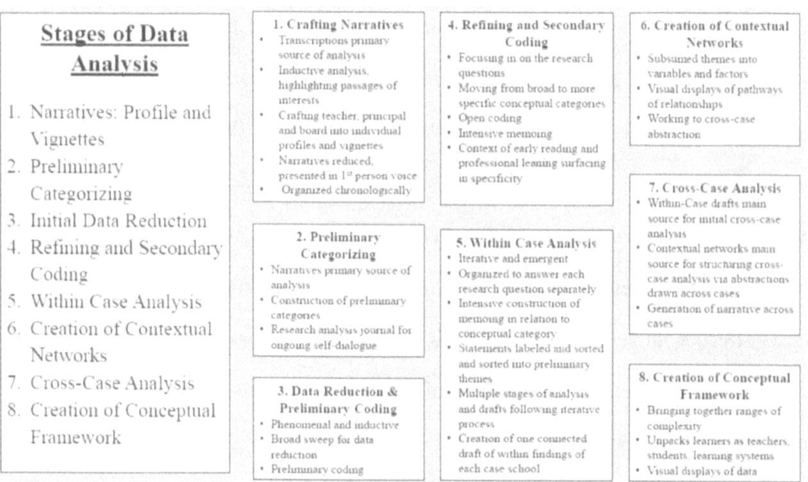

According to Attinasi, Jr. (1991), a phenomenological inductive process is a place for concept

and hypotheses to emerge out of the "the informants' lives reported in the interviews [with the]

induction process … constrained only by the research perspective" (p. 3). In embracing the messiness of complexivist research methodology, the analysis occurs across eight stages involving multiple deep steps of exploration and continuous data reduction (e.g., Koopmans, 2014). Seidman's in-depth three-series interview process structures the deep probe into the lived experiences of participants. Following the suggestion of grounded theorists Glaser and Strauss (1999), I initiated a process of memoing in each stage of analysis to streamline the depth of data into their working categories. Initially, I transcribed the 24 interviews verbatim using Microsoft Word 2010. Transcripts were transferred to QSR International's NVivo 11 software for preliminary analysis and conceptual sorting before transferring the data back to MS Word for most of the heavy work in the data analysis. Seidman's (2013) processes for data analysis of my in-depth three-series interviews with teachers broadly structure my data analysis: (a) studying, reducing and analyzing the text, (b) creating participant profiles and themes, and (c) making and analyzing thematic connections. Narratives in the form of extensive profiles for teachers and shorter vignettes for principals are the initial reductions of the data. They occur inductively rather than deductively by allowing the first stage of analysis to provide emerging conceptual categories (see Appendix J-O for the full profiles of the teacher participants).

Across all interview data, Merriam's (2009) guidelines for category management and coding construction influence the analyses. Based on the reasoning and work of Seidman (2013), the teacher and principal participants' profiles develop from the interviews and portray a layered narrative for structuring perspectives across levels. Following the profile construction, the narratives were then used as the primary resources in the categorizing and coding processes for the within-case analyses (teachers and principals).

The construction of contextual networks arose after multiple in-depth within-case analysis and the resulting within-case findings. These networks provide a way to open up cross-case abstractions. Creating contextual networks is a unique process, and so there is a description of this in more depth below. Following multiple iterations of in-depth within-case analyses, I borrow from Miles and Huberman's process of creating visual displays of causal networks to demonstrate teachers' perceptions of how their early reading professional learning influences their practice and how contextual variables interact to influence their teaching and learning and their students' early reading achievement (Miles et al., 2014). From a complexity perspective, this process provides a representation of contextual variables and factors interplaying within the scope of the research questions and continues to move from sequentially answering each question separately while developing an understanding of how these relationships appear across cases (Davis & Sumara, 2006). To create these networks, this part of the analysis follows Miles and Huberman's suggestion for listing antecedent and mediating variables (1 antecedent through multiple mediating) and outcomes. It is an iterative phase of analysis, and an example of this process is in Appendix U.

Connecting to this stage of analysis are three figures to display how themes subsume into the different contextual variables of each contextual network. There is a more in-depth discussion on this, along with visual displays in the next chapter. This stage of analysis takes its formal shape through seeking to identify the variables and factors linking to perceptions of influential early reading professional learning and also how they link to participants' perceptions of what influences their students' reading success; in the barest sense, this is exactly what the research questions seek to answer. Miles and Huberman speak to the value of narratives as secondary but essential aspects of data analysis and representation. Following the creation of pathways, are narratives for each contextual network that synthesize the within-case findings of contextual variables at play within each participant's teaching lives at each of their schools. In doing this, I distil the themes derived in the process of within-case analysis and move from case narratives to contextual networks that portray the movement of and between the variables participants perceive influential to early reading professional learning and student reading success. The contextual networks then contribute to the generation of the themes in the cross-case findings.

Following within- and cross-case analyses and the writing of these findings is one more step of analysis: the development of a conceptual framework. This conceptual framework emerges from answering each research question using content from each theme in each stage of analysis. The within- and cross-case themes are colour-coded to relate resemblances with each theme numbered in tables for answering the research questions. These tables were organized by research question to understand (a) how contextual variables at the school, board, and other levels are perceived to influence the planning, delivery and uptake of early reading professional learning and (b) how teachers perceive the relationship between their professional learning, their early reading practices and student reading outcomes. These tables are used to generate how varying factors influence the main variables (teachers, students, school system). Taken together, notions of self, the nature of teaching, empathy to the nature of the student, and the relationship to learning in a learning system paves a path for conceptualizing the complex interrelationships relating to teacher learning and change and student achievement (see Appendix U for more details and a lengthier explanation of my process of data analysis).

Trustworthiness

Credibility, transferability, dependability and confirmability are four major criteria qualitative researchers have expectations to meet for ensuring trustworthiness of their research (Guba & Lincoln, 1982; Nowell et al., 2017).

Credibility

Credibility relates to the interpretation of the data representing realities expressed in the perspectives of the study's participants. Guba and Lincoln (1982) emphasize that "the crucial question for the naturalist becomes, 'Do the data sources (most often human) find the inquirer's analysis, formulation, and interpretations to be credible (believable)?'" (p. 246). They list six ways to help safeguard against the loss of credibility (prolonged engagement at a site, persistent observation, peer debriefing, triangulation, referential adequacy of materials and member checks), and these are all met to varying degrees in this study. Interviews with principal and school board participants gather their perspectives around the context of the school and board and enhance an understanding of teacher learning and early reading instruction in schools and across the board. Principal and board participant data provide context and a part of triangulation through collecting their different perspectives. Each of these participants and the six teacher participants were provided with their transcripts to review and make changes if they perceived any misrepresentation; no participant made changes. The structure of Seidman's (2006) three-series interview process with teachers provides the opportunity for prolonged engagement and persistent observation, the first two ways to safeguard credibility as listed by Guba and Lincoln (1982).

> The three-interview structure incorporates features that enhance the accomplishment of validity. It places participants' comments in context. It encourages interviewing participants over the course of 1 to 3 weeks to account for idiosyncratic days and to check for the internal consistency of what they say. Furthermore, by interviewing a number of participants, we can connect their experiences and check the comments of one participant against those of others. Finally, the goal of the process is to understand how our

participants understand and make meaning of the experience. If the interview structure
works to allow them to make sense to themselves as well to the interviewer, then it has
gone a long way toward validity. (Seidman, 2006, p. 24)

Peer debriefing, the third way to deliver credibility, occurs in the conversations with my
doctoral supervisor throughout this research journey (electronically and face to face). Regular
meetings throughout each stage of research provided the "opportunity to test [my] growing
insights against those of uninvolved peers, to receive advice about important methodological
steps in the emergent design … and to discharge personal feelings, anxieties and stresses that
otherwise might affect the inquiry adversely" (Guba & Lincoln, 1982, p. 247). This ongoing
dialogue and the inclusion of perspectives from a range of participants across the various
contexts of the case schools contribute to triangulation.

Referential adequacy materials, the fifth element to ensuring credibility Guba and
Lincoln (1982) espouse, reflects in my adherence to maintaining and storing materials beyond
the data that I collected for analysis. An electronic research journal of over 150 mostly single-
spaced pages was maintained throughout the analysis and writing up of the findings. A collection
of paper notebooks contain detailed notes of my research process and early drafts and
conceptualizations of the visual representations in the book (e.g., emerging thematic
relationships and connections, contextual pathways, conceptual framework). Stages of analysis
are documented on paper and these files are organized and securely stored away; available for
demonstrating the direction of this part of my research. The final element to establish credibility
listed by Guba and Lincoln (1982) is member checking. I shared transcripts with all participants
with offers to provide feedback, and none did. The nature of the analysis is emergent, and
analysis comprises a lengthy process of varying stages, as shown in Figure 6 above. Using

crafted profiles based on the first-person voice of my participants and ensuring regular communication with my doctoral supervisor helps me provide findings deeply based on the teacher participants' lived experiences in their case schools. Additionally, I will share the findings with the case school board and case schools at a time and in a format that is convenient for them.

Transferability

The second criteria for establishing trustworthiness is transferability (Guba & Lincoln, 1982). The two elements to ensuring that the data and findings can be transferable are theoretical/purposeful sampling and findings presented through thick description. My research design utilizes purposeful sampling. Context is provided by administrators and board level reading experts, along with the teachers. Purposeful sampling of teachers focuses on early reading teachers, and the case schools are selected based on their demographic differences and representation of other schools in their school board (rural, urban, suburban). Thick description of the teacher data is a result of my in-depth three-series interviews. First, all interviews were organized into first person narratives. The teacher narratives are in Appendix J-O for transparency and because they provide an intimate portrayal of the lived experiences of these teachers as readers and as professional practitioners. These narratives provide "enough context, first, to impart a vicarious experience of it, and, second, to facilitate judgements about the extent to which working hypotheses from that context might be transferable to a second and similar context" (Guba & Lincoln, 1982, p. 248). Within-case findings provide a depth and range of direct quotations from the participants so that their voices resonate throughout the findings.

Dependability

The concept of dependability relates to stability of the study so that the methods that are used would, if used by another researcher, generate similarly expressed interpretations (Guba & Lincoln, 1982). Three strategies Guba and Lincoln suggest for establishing dependability are, (a) the use of overlap methods, (b) stepwise replication, and (c) the dependability audit. Though interviewing is the vehicle for data collection for all participants, I employ various interview styles depending on the type of participant. Step-wise replication occurs through the structure of my three-series interviews with teachers. Setting the series of interviews over three weeks for teacher participants ensures consistency in perspectives and adds depth to the intentions of each of the interviews. A dependability audit occurs in the professional conversations with my doctoral supervisor: throughout this research, the details and the direction of this research are transparently and thoughtfully shared. Further, I document the research process, how I organize and store data, the stages of analysis, and writing of findings in each stage of this research journey.

Confirmability

The final criteria Guba and Lincoln (1982) list for ensuring trustworthiness is confirmability: "The onus of objectivity, ought, therefore to be removed from the inquirer and placed on data; it is not the inquirer's certifiability we are interested in but the confirmation of the data" (p. 247). Achievement of confirmability occurs through triangulation, practicing reflexivity and an audit. Triangulation, mentioned earlier, is ensured by including different perspectives (varying teachers, principals and board-level reading experts), using different interview structures (three-series, in-depth with teachers, focused semi-structured with principals and board participants) and having a diversity of case schools (rural, urban and suburban). Keeping a research journal to challenge my underlying epistemologies, to self-dialogue on the emerging themes in my research, and as a place to document the development of my

interpretations is maintained throughout my research, but particularly during my analysis and

write up of the findings. In terms of a confirmability audit I speak above to the maintenance,

orderliness and storage of the process of my research (e.g., labeled, time-stamped and organized

according to aspects within each stage of my research).

Chapter 4 – Findings

The purpose of this research project is to explore, interpret and understand how early reading teachers perceive that their early reading professional learning influences their reading instruction and improves their students' reading achievement and, secondly how contextual variables from varying levels interact and are perceived to influence teachers' early reading professional learning experiences. This chapter presents key findings from the analysis of 24 interviews of teachers, principals and board-level reading specialists. A qualitative case study approach at the school and board level shows the contextual uniquenesses and abstractions across cases. In relation to teaching, learning and early reading instruction, this chapter begins by presenting a brief description of contextually relevant information regarding the province where this research took place. This context mostly derives from the perspectives of the school board participants. This information provides an introductory contextual setting that leads into the contextual descriptions of each of the case schools. To answer the research questions, I then proceed through the within-case findings, contextual pathways and cross-case findings before presenting the conceptual framework generated from the deep stages of analyses.

Provincial Context and the School Board

This section illustrates the elements related to early reading instruction and professional learning through the perspective of the board-level participants, all who have extensive early reading knowledge and experience. This descriptive account merges the perspectives of the three school board early reading participants to provide a general sense of reading instruction within the structures of the provincial education plan as well as the collective sense of early reading from a board-level perspective. Next, a contextual snapshot of professional learning in general and reading professional learning, more specifically, is presented based on my interviews with SB1, SB2 and SB3.

The Provincial Education Plan

As described in Chapter 2, the implementation of the current provincial Education Plan at the beginning of the twenty-first century shifted the curriculum from traditional outcomes towards competency-based learning. The traditional single grade system became cycles comprising two grades. There are three cycles in elementary schools (Cycle 1: grades 1 & 2; Cycle 2: grades 3 & 4; Cycle 3: grades 5 & 6). At the board level, SB3 remarks that, for teachers, the competency-based expectations have been "really hard for them," and that she and SB1 currently worked with teachers to help simplify an understanding of the curriculum and how literacy instruction looks across a cycle and, more specifically, within grades. In theory, SB3 notes that it is a framework of a curriculum for connecting community needs, resources and context that make learning more meaningful than the more traditional curriculum before it. She suggests that the education plan is a modern curriculum that provides space for teachers to be flexible in how they teach, which is not prescriptive and is cross-curricular.

However, the board participants perceive that the plan lacks specific learning outcomes, making it hard for teachers to assess specific skill development. SB1 remarks, "The QEP when it came … was a curriculum ahead of its time … and the problem is … it's so difficult to apply to the classroom. There's so much wiggle room that people don't even know what they should be teaching and because of that it's very hard to … build on skills." SB2 believes the plan ambiguously states outcomes for students and that its competency-based framework is somewhat unclear for teachers to help them design their reading programs. There have been changes in how reporting student achievement is done and that the changes do not necessarily align with the competency-based nature of learning in cycles. These changes appear inconsistent with ministry expected products for teaching, evaluation of progress, and formal reporting expectations represented in the provincial curriculum. For example, initially, the education plan required teachers to report achievement on a four-point scale from "not meeting" to "exceeding." Enough parents complained that this was not transparent and specific, "and so the government came and said we need to be clear and so now you are going to start using percentages … they don't match … and that's just the way it is and so yeah we've got a conflict going on here of how things are supposed to be going and then how we're going to assess it but they don't match."

The board participants clearly empathize with the challenges teachers face in terms of the unease and conflicting expectations they are experiencing in teaching within the current curriculum. A lack of specificity within the curriculum on the early reading skills to teach and equitably assess reading achievement appears to provoke some discord in how language arts are understood and approached.

Professional Learning Structures

A sense of the nature of professional learning in the board can be understood by how SB1, SB2, and SB3 perceive how professional learning subject and availability can vary in terms of access, topic, and style of delivery. SB1 considers this school board to be small in population. She also suggests that this board allots more funding for academic consultants than any other board in the province with the current board participants in this study being highly experienced and knowledgeable of early reading programming and instructional strategies. It also appears that the nature of the consultants' roles has changed. Whereas consultants used to be quite field-based, SB3 believes that the board would like them to work centrally and providing resources for teachers instead. While geographically large, similar to the size of Nova Scotia, its overall population is considered small and all board participants mention this limits the amount and variety of professional learning that the board can offer. It is challenging for consultants to go out and see all the schools. Likewise, as identified by OT1, OT2 and OP, it is difficult for many teachers who are distant from the board office to participate in person in centralized professional learning. In this scenario, it appears that schools that are remote from the central board are less likely to visit or receive the professional learning available to other schools. Partly for reasons of equity, SB3 is a proponent for literacy experts attending schools and notes that working with teachers over a few days is the most effective way to influence teaching.

The structure and topics of board-supported professional learning initiatives have gone through varying transformations over the last couple of decades. Two of the previous board-wide learning initiatives focused on instructional intelligence and a cooperative learning team-building initiative. SB3 remarks on how the board integrated the two topics and that this was "hugely positive and really good [professional development]" across the school board. She notes that some workshop opportunities for learning about and implementing the team-building initiative for teachers new to the board were available. However, this initiative was mainly focused on individual schools and not often offered across schools by the board as professional learning. SB3 notes how following the widespread focus on instructional intelligence and cooperative team-building, the structure of much board provided professional learning moved to learning networks and a "train the trainer" model.

For early literacy professional learning, this comprises a focus on developing networks of literacy leaders who bring knowledge and resources back to their schools following workshops at the board. SB3 finds this a useful and engaging experience for literacy leads from each school. However, the same teachers will often represent their schools, excluding many teachers from having first-hand learning opportunities: "When you go back and you just tell people what you did it's not the same, you don't own it" (SB3). She finds that this network does not work well for teachers in more distant locations with SB3 suggesting that virtual conferencing lacks the quality and social intimacy of in-person experiences.

Change of directors who oversee literacy and professional learning, maternity leaves by literacy leaders, and union directives also appear to influence the flow and opportunities of professional learning. SB2 and SB3 both speak about the change in direction and directives that occur when new directors take control of structuring professional learning or board learning goals. Short terms leaves also influence and disrupt the flow of different learning initiatives or ways of doing things. For example, recently, SB1 returned from a two-year maternity leave. Initiatives had finished up during her leave, and nothing new had begun while she was gone. SB1 was about to begin her second maternity leave, likely causing another disruption. On this, SB3 remarks how hard this lack of continuity is on longer-range planning.

Scheduling and timing are critical barriers to getting teachers together. For example, I was told there are union protections or priorities that influence teachers' opportunities to attend board provided professional learning. SB1 explains that the union mandates professional learning to occur within work hours. For the few teachers who come from far away, they are often late and leave early. There are also challenges for teachers to attend learning outside of their schools, although there are quite a few PD days during the calendar year. Says SB1,

> the unions have negotiated that teachers shouldn't be pulled from their schools too often
>
> … Like they're supposed to be able to use those PD days to get work done in the
>
> classroom and because teachers' PD days are connected with bigger conferences

happening twice a year it is hard to create times for all teachers to meet on professional

learning days.

In this section, environmental context, professional learning design, personnel changes, and

varying controls over how professional learning can be utilized indicate the constant ripples of

change interacting upon and influencing the products of professional learning experiences. Next

comes an opening up of the perspectives on reading in general and in the board, offering a

portrayal of the current reading environment through the eyes of the three school board reading

consultants.

Reading Professional Learning

Currently, SB1 sees the initiatives in early literacy focusing on narrative writing and

response because these are the "two components on our provincial exams." SB2 perceives that

professional learning focuses on preparing grade 4 and 6 students for provincial exams. She

notes the board's past push on early literacy, but that lately, the focus is on boosting literacy and

numeracy assessment scores of the grade 6 students.

In terms of learning strategies in early reading professional learning, SB2 perceives that

the board is not helping early reading teachers develop their reading instruction: "I don't think

there's been anything for years." In response to a question on availability of training, SB1

remarks, "I'm going to be really honest and say none [and] we have no formal training that we

do through the school board in early reading. I mean we offer PD workshops where we look at

different areas in reading and writing but we don't sort of follow any kind of program."

Though there is no mandated early reading professional learning and instructional

program, there are two board supported reading initiatives occurring over the past ten years that

the consultants consider useful for providing a balanced literacy approach. One program relates to learning how to implement a reading program comprising a range of learning centres such as guided reading, and reading and writing to self and peers. The second partially resembles a code-based program where students learn nonfiction vocabulary while developing some phonological awareness. SB1 mentions that many teachers implement these initiatives in varying forms in classrooms throughout schools and across the board. There is an expectation that teachers who receive training in one or both of these initiatives are implementing them similarly board-wide. However, the consultants also perceive that resources from Pinterest, Teachers Pay Teachers and other Google searches embed themselves into teachers' early reading instruction.

Perspectives on the State of Early Reading Instruction

The school board tends to encourage a balanced literacy approach to Language Arts. Over approximately the past ten years, the two board supported reading initiatives and teacher learning opportunities reflect this. As noted above, one initiative focuses on setting up literacy stations, and the other initiative is a phonics-based vocabulary program. Though there is a general perception that most teachers provide a balanced approach, implementation varies quite a bit across classrooms. SB1 suggests that only about 20% of the teachers who commit to the board supported reading initiatives are implementing them in a student-centred fashion, the way these programs are intended. According to SB1, one reason for this range of variability is teacher changeover: "I'm sure half the teachers I've worked with are somewhere else now." SB1 and SB3 also believe that many teachers do not include significant parts of the program in their teaching by either skipping elements or trying to implement the whole approach too soon rather than setting up routines necessary to help students understand how to work independently at the different literacy stations.

Teaching, Learning and Transitions in Grade 1

The board participants perceive that teaching grade 1 is one of the most challenging teaching assignments. Transitioning from a play-based environment into an environment of standardized reading and assessment is tough on students, and SB1 believes that teachers' needs are not reasonably considered: "teachers aren't being supported the way they need to be and that's because assessment drives pedagogy." SB3 sees teachers stuck in the pressures of getting kids to move levels and that often this excludes students' social and emotional development: "I find that's the hard part for them. I think they feel they have an obligation to teach that kid to read and this is how it's done … [for teachers] I think it's feeling confident enough to say 'I'm going to take my time here.'" SB2 calls for a rethinking of the role of assessment and believes there needs to be "almost an enriched kindergarten program for the grade 1s coming in and forget about the testing … and see where they are in January, February" instead of assessing reading ability in October on standardized reading diagnostics. Continuing, she speaks about the limited scope of reading achievement and the limiting role that assessment appears to play in teaching:

> I think teachers are teaching to the test … And that it follows under the umbrella of the school board … it's all results driven which every school board is … So yes it's the improvement of the literacy but at every school it's more finite and every teacher it's more finite so it just trickles down but it's very general you know, it's very general but there's no specific direction.

A sense of pessimism continues with SB1 remarking how even though some principals are implementing more social and emotional learning opportunities in their schools, it is not going to "change anything in terms of culture" because teachers are locked into a milieu of assessment and measuring student achievement by norms that are not currently relatable for many students.

All three board participants remark that many students in this board from rural and urban areas lack a literature-rich background and do not have the proficient language comprehension and vocabulary skills that are important predictors of learning to read. SB3 sees teachers too reliant on levelled texts, and students are learning skills, but often in their early years their vocabulary is not there and that "they're going to have trouble holding that language pattern in their head." She suggests incorporating more play into early literacy instruction by using poetry and chants as engaging settings to help students build their vocabulary banks.

SB2 suggests that there is not enough early intervention happening in kindergarten and grade 1. Too often, students receive interventions too late, and early intervention would help many students learn helpful strategies to compensate or overcome their struggles. SB3 also believes that understanding the nature of why different students struggle with reading is not well understood by teachers and SB2 perceives that currently, teachers are too quick to suggest a learning disability or to request outside intervention because they are not confident in their ability to help students who are struggling to learn. There is a perception that there is not much assistance from the board to help teachers develop more effective ways to meet struggling readers' needs with SB1 lamenting:

> teachers are very overwhelmed … they have lots of kids in their classroom and I think differentiating for different learners is really an area of weakness in a lot of our classes … it gets compacted with every year that goes by … I think we know what we need to do. I don't think we have as many teachers that are skillfully doing it.

Case Schools: General Context and Description

This section describes the context of the three case schools beginning with Blending

Elementary, moving to Outer Elementary and ending with Central Elementary School. A general

overview for each school is in Table 3 below. The structure of the case narratives provides a

short description of the school context and includes information from each case school's

participants. Included in each narrative, a section on school leadership represents the principal's

role in school professional learning, opening up the general context of each school and situating

each principal into the school's local interactions.

Table 3

Overview of School Demographics

School Name and Abbrev.	Type of Pop. Density	Grades	Student Enrol.	Majority First Language Spoken at Home	Teachers on staff	Classes with Grade 1	Classes with Grade 2
Blending (BES)	Suburban	PK-6	330	English	21	2	2
Outer (OES)	Rural	PK-8	100	French	7	1	1
Central (CES)	Urban	K-6	550	English	35	5	5

Blending Elementary School

Blending Elementary School has bright open rooms and a large fenced-off schoolyard

with a modern playground. Blending is one of several villages comprising a small municipality

with an approximate population of 7500. Predominantly an English speaking community,

Blending Elementary School services a population of about 330 students from pre-kindergarten

to grade 6. There are about 20 full-time teachers and a school principal. Blending and a few other

villages very close by feed into BES. The population comprises a mixed socioeconomic range.

Many of Blending's working population commute to one of the two larger cities, approximately 25-45 minutes south. Other adults comprise a population who work out of their homes, work in agriculture, in the trades, or hospitality businesses. There are ski hills close by, and the scenic outdoor opportunities, as well as Blending's proximity to urban populations, draws tourists throughout the year. Participants describe the school community as a staff of educators who cooperate and collaborate; and that a core group of teachers remain and help shape a general perception that teachers are on board with and available to each other.

Context of Leadership at BES. BP spoke to me about how she sees herself as a school leader with over 20 elementary teachers. She remarks that she expects teachers to self-initiate their learning processes and that she is always willing to help out those who strive to professionally develop. She says that "there's a cultural expectation to keep learning" at Blending and that, for the most part, teachers display a growth mindset. She enacts a sense of distributed leadership in which everyone acts as a team, and experts within the school need to be tapped into. Success rates are of obvious importance, and BP strives to hire teachers who fit in with her instructional beliefs: "You have to really find somebody that fits your way of thinking and not everyone does … I am looking for somebody who works collaboratively, who wants to share, who wants to learn, who wants to grow as a professional and who wants to focus on teaching and learning, not on you know, all the minutiae that comes up."

BP perceives herself as a flexible administrator, teacher-centred in her leadership approach, but firm in her expectations that teachers are accountable for their self-development. She sees herself supporting her teachers' professional learning needs in three overarching ways: (a) providing resources to attend learning experiences or suggesting other classrooms in the board to scaffold their practice; (b) bringing in consultants to help with teachers who are struggling in certain areas; and (c) providing in-school professional growth topics for her staff that she believes will contribute to whole school success (e.g., recent universal design for learning sessions and ways for making learning a visible and student-centred process).

The school's professional image is critical to her and so she has had to intervene on how and when the school schedules their four floater professional learning days. She will hold back introducing new professional learning topics (up to a year or more) until she is confident the teaching staff will be receptive. Organizationally, she prefers teachers to work in pods and cycle teams on common professional development topics relevant to their teaching grades and responsibilities. BP views the main barriers to teachers obtaining effective professional learning often are the teachers themselves. Too often, when teachers miss an opportunity, the teacher has decided that they do not have the time or professional willingness to take on new learning challenges.

BP is interested in John Hattie's visible learning research and believes that this has been a turning point for her in how she conceptualizes instructional dialogue in and as student-centred learning. She reserves some of the school's professional learning times for working with teachers to develop an understanding of the relationship between learning intentions and success criteria, and in bringing the students more explicitly into this process. In terms of literacy and, more specifically, reading instruction, she speaks about the shift from earlier in her career where whole language was the philosophical approach in universities and schools. With the recent board reading initiatives she believes that opportunities for more students to learn at their levels are available as the approach to reading instruction that is currently supported by the board "allows you to have the kids autonomously walk through at their level ... with real meaningful texts but then the teacher can also have that guided reading group going on all the time."

Outer Elementary School is located in Extére, a small, rural, and predominantly (90%) French-speaking community. Extére has a population of approximately 600 that contributes to OES' student body along with pulling students in from eleven rural municipalities located nearby. Extére's leading source of employment is forestry. Tourism is a much smaller but still significant source of the region's economy as Extére's proximity to rivers, lakes, and mountains makes it an idyllic location for outdoor adventurists. However, Extére is not an economically thriving area. Ranked nine on a 10 point socioeconomic scale, OES is designated as a low socioeconomic school that receives extra resources from a consultant with expertise in helping teachers in low-income schools. Having 90% of students with French as their first language and a high rate of poverty creates unique challenges for teachers and students. However, it is the high level of poverty that OP sees affecting her students' learning upon entry into schooling remarking,

> what you're seeing is those students even if they're first language speakers, they're coming to school with a third of the vocabulary of an average child whose been read to nightly, who has not been babysat by a screen for a majority of their time, has had time to interact and develop vocabulary and it puts them at a loss right from the very get go … And in our situation their vocabulary in their mother tongue is weak and then we're throwing in an extra language … those first years are spent catching up on a lot of oral language before we even get to reading and writing things.

OP spoke about how she and the staff continuously try to build relationships with students' parents. Many families have a long history of living in this area, and the notion of schooling is not a positive one for many parents. Trying to bring parents in for workshops on how to handle homework and sharing reading tips for reading with their children is challenging. OP speaks of trying to create an inclusive environment for parents by "spend[ing] a lot of time looking at what those parents experienced as students … so just getting them back through the door is the challenge at the beginning … it's constantly about trying to get them comfortable in the building and having that open-door policy." A cycle of sensitivity and negativity is trying to be eroded by OP and her school team. Building better relationships between the school and home are essential for bridging the learning gaps their students face.

There are less than ten teachers and one administrator at OES. The principal, OP, has a forty percent teaching assignment, and the annual student population hovers close to 100-comprising four-year-old kindergartners and up to grade seven students. Teaching assistants and supply teachers are not necessarily professionally trained. Several teachers live an hour or more from Extére, and the central board office is a 115 km drive from OES, which can make it challenging to attend professional learning at the board. OP remarked that staffing had undergone much change with teachers moving to more centrally located schools making it a challenge to develop their school identity. However, a sense of stability with a current core group of teachers who have remained is emerging. Extére's rurality is associated with poor internet connectivity, and this is perceived to negatively impact web conferencing and the use of assistive technology in the classroom. OT1 sums this up: "Like technology here … especially in this rural area where we're not serviced properly and we lose power often you can't depend on it … You know probably 30% of the time our internet's down … So while technology is awesome it's almost twice as much work sifting through things and you anticipating the worst."

The Context of Leadership and Learning at OES. OP appears as a leader who seeks change through her close connections and understanding of the school community's needs. In her administrative role, she spends as much time as she can to visit classrooms. She believes this creates close connections with students and teachers and acknowledges how her relationships influence her thinking towards professional learning and school success. OES' small population also requires that two-fifths of her position is teaching. It provides her opportunities to work one on one, in small groups and classrooms with a range of students. A principal with this teaching responsibility has a viewpoint for closely observing and understanding the most current student needs and sees herself "as someone who can take the vision and help carry it forward."

OP's accumulating professional experience and maturity as the school principal also appears influential upon the structural supports to school-based professional learning. Her first few years as the principal at OES were overwhelming due to engaging in too many staff professional learning initiatives. Following her first few years, she now perceives that it is healthier to reject much of what is available and develop professional learning through a collaborative process with teachers to determine their school's priority needs. Part of this she believes is by helping to instil a sense of passion for reading within students: "we always come back to the point they never lose the love for good literature and that they learn to not look at reading as a chore but something that you do that opens up the world to you." She also understands the challenge of instilling this passion when many of the students at OES have limited English language backgrounds and experiences and, therefore, a low English vocabulary bank necessary to decode, predict and comprehend and meet board reading standards for English Language Arts. OP's perspective on the challenges struggling readers face illustrates the complexities that exist as the staff considers what early reading professional learning can be meaningful within their school's context. Together it appears these perspectives intertwine in the direction of school-based professional learning goals that focus on vocabulary development and visible learning (externalizing perspectives on metacognition and learning needs).

OP senses that there is a growing connection between the staff and how they take on professional learning. Reverberations emanate from student and school needs, travel between staff by bridging learning to classroom practices and then regularly looping back into staff discussions. Learning occurs in staff meetings between triads planning their classroom strategies and in the halls and classes, during teaching and over lunch. Speaking about the ongoing nature of their learning OP says: "we have homework between our PD sessions so they will naturally go to groups of two and three and work on something and come back. It's nice to see." OES' school growth planning and professional learning appear to be a collaborative process directly based on their students' needs and OP's leadership philosophy: "As soon as I step out of that classroom I worry that I'm going to lose that perspective … I worry that losing that perspective will make me a distant administrator rather than a hands-on one, and I still need to be there for my own professional learning."

Central Elementary School

Comprising junior kindergarten to grade 6 students in either the English or French immersion stream, CES is located in the urban heart of a large city with a population of approximately 330, 000 and is one kilometre north and divided by a provincial line from a larger Canadian city with a population of nearly 1, 000 000. Though CES is a downtown urban school, the students bus in from as far as a ten-kilometre radius, not walking or biking because these students do not live in the middle and upper-class neighbourhoods their school is tucked away in. It is an old school nearing its 60th anniversary. CES is low on the socioeconomic factor, an eight out ten, and receives additional grants from the Ministry of Education. Though it is predominantly a low socioeconomic school, it includes many students from professional families who are more prosperous than the school's rating suggests.

CES is a community-based school. The principal believes this is largely because many of the students do not live close to each other and are bussed to the school, creating the school as the natural hub that connects parents, children and school staff. CP dedicates his time to create a community for families at the school:

> [CES has] had things that other schools, that probably wouldn't dream of, cultural activities, Pow Wows, weekend events that are constantly going on to bring parents and families and teachers into the school whether it's to play together, whether it's to celebrate together, have fun together, learn together. (CP)

CES provides a lengthy list of services as a community learning centre such as parenting, anti-bullying and business workshops, a range of clubs, breakfast and lunch services for many students, evening sports nights, and monthly activity afternoons. CP exemplifies CES' community-minded philosophy in how he views the connection between all members of the school community:

> When there is that interplay or interchange of parents and teachers and kids, the kids see their teachers differently, the parents see the teachers differently and then most importantly the teachers see the kids differently ... They see their strengths and when they take that back into the classroom the teaching and learning and the behaviour is all going to be better ... So it's just a different, it's way of looking at things ... Then you have to organize differently. There's a creativity to organizing a school that can't be traditional because if you get stuck in the same 19th century way of organizing a school the kids

aren't going to be learning the way they should be learning today. It's maybe a touch unique but that's what the school needs I think.

A staff of 55 educators, more than 500 students and their parents who regularly volunteer and take part in many of the services offered to families, CES is literally bursting at its seams. Previously under the threat of being underpopulated, CES has doubled in population over the past five years and now is in danger of being overcrowded. With classes over the size and composition limits, not enough books for student home reading, resource teachers without a home base for working with struggling students, a library regularly used for a grade 6 math class instead of enriching-literature opportunities, CT1 remarks that

> it's great to be an awesome school, it's not great to be an awesome school stuffed to the rafters … you can really feel it in this school from day to day. You can really feel the energy, feel the crampness … some of the classes are really big and really hard to teach and teachers are burning out and … I don't think it's a great scenario. I don't think the size [population] of it is a problem, I think the [principal] has to solve it by making more space and bringing the class sizes [down] so the teachers aren't taxed so hard.

The popularity of CES is largely attributable to its commitment to the community and other attractions to the area that pulls a range of families in for a limited time and this contributes to a transitory environment. Its central location, being adjacent to a sizeable Canadian city in another province, proximity to housing for military families, opportunities for cheaper housing within the catchment area, and its closeness to an adult education centre combine to create a current of children in and out of the school throughout the year. Many of the students making up the transitory population are children who have moved multiple times, are in and out of CES as their parents come from northern communities to complete a course at the adult education centre, move back upon completion, and then return later repeating this process. In other instances, families use the catchment area as a temporary stopover before moving into more desirable locations.

Part of understanding the contextual uniqueness of CES is that it provides two streams of instruction for students, and these streams appear to play a central role in how classes are composed. There is an English stream and a French immersion stream to compete with other schools in the vicinity. CP speaking about the importance of offering French immersion at CES:

> if you don't offer a 50-50 immersion program many parents are not going to select your school because they want as much French as possible and ideally [for] professionals or government workers they know you need as much English as French and if you just go to the French school you can't get the English.

At CES, the principal, with consultation from the teachers, arranges classes based on students' overall readiness (academic and social and emotional). Based on their readiness, students are placed into one of four quadrants. In the principal's perspective more than half of the current students are not ready for learning in the traditional classroom setting, so most classes are composed of students at comparable readiness levels. However, according to CT1, this can be a challenging undertaking for teachers provided with academic and socially emotionally struggling students. Of note, the higher academically and socially ready students are more likely to be in the French immersion stream, less ready students tend to be placed in the English stream. Connecting to this, it appears that immersion teachers often request that their struggling students be placed into the English stream. CT2 conveying a tone of desperation was emphatic about a double standard impacting lower level students in the French immersion stream:

you can't kick them out of grade one French immersion and that's what happened …

We're trying to get away from that but it happened again because they can't learn French

immersion because they're really weak. So where do I throw my kids if they can't read

English … You don't get to pick who you teach. You get [your] child, you adapt things

and teach them to who they are.

This sentiment echoes in the words of CT1, teaching in the French immersion stream during this

study with a relatively high achieving group of students, as she reflects on ideals she perceives

missing within the French side of schooling:

There's been a lot of dissatisfaction in the local French schools and I think that their

language instruction from what I gather is very different. It seems the way kids learn in

the French schools is very different from the way we teach things in the English system

… You know how we differentiate a lot right, we've been even going through that with

our French teachers here. Like they're "well they can't pass," and we're "so bring it

down for those students," you know. They kind of teach [at] one level whereas we're all

taught, which I think is really important, to differentiate.

The Context of Leadership at Central Elementary School. At CES, CP holds

particular beliefs about the critical attributes primary grade (and particularly grade one) teachers

should possess. He also believes that it is his responsibility to help transform CES into the

community hub. Noting that students are bused in from across a ten-kilometre urban radius, he

seeks to create opportunities to bring teachers, parents and students together and soften the

boundaries differentiating those with different roles. Capitalizing on these opportunities is one

way that he sees that school communities can organize themselves to connect to students' social

and emotional attachment needs. Estimating approximately 40% of CES students are not reading at board-level grade one expectations, he believes that early literacy needs to be reconceptualized by including more opportunities that develop teachers' social and emotional competencies.

CP describes how he prioritizes pedagogy that places students' learning experiences within their individual social and emotional needs as the new lens for "understanding the preconditions getting to that point" of detachment. Additionally, he values how a uniform approach to early reading instruction can create a positive learning environment and would like all the early reading teachers at CES to actively practice the board supported programs. CP recognizes grade one as the most critical year for most students. In his words, the grade one teacher

> has to be at the top of their game … taking it to another level, who's got the capacity,
> relationship and even philosophical to a degree and can see a different way of doing
> things, could collaborate with kindergarten teachers, could find a way to make room in
> their grade one class for what the kindergarten teachers are doing with structured play
> and movements and taking less of a focus on what the curriculum is expecting them to
> do.

Noting that "teachers there have to have it at all levels" he also realizes that he cannot place all he considers to be the "best" teachers in first grade. He also perceives that it is his responsibility to try and place teachers in the grades that interest and motivate them. CP believes that healthy schools are places where students and teachers are put into positions to succeed.

In timing professional learning opportunities, CP understands that it is important to consider balancing professional learning with teachers' personal and professional obligations in mind. In the school board, many professional development days are close to other busy times in the year, and he mentions how sensitive he has to be towards the rhythm of the calendar. At this stage of his career as an elementary school principal, he believes that he recognizes the correct times to push his views so that the teachers are willing and engaged learners. One way he sees this happening is by creating a network of learning teams within the school. By embedding teams with teacher leaders who receive training from consultants and experts, CP believes expertise can be generated and sustained at a school level. Coordinating this element within professional learning involves foresight and patience. He also believes that a learning network's long-term sustainability is dependent on several factors that are beyond the expert's and teacher leader's control: commitment to attend team meetings, individual initiative and desire to learn, an understanding of how learning teams can enhance the students' learning environment, and the perspective or belief of some teachers that his direction in terms of professional learning is an imposition.

CES has its unique context among the schools in this case study: overcrowding, a transient nature generally comprising those who need stability in a nurturing environment, tensions and challenges in having two streams of instruction, and its role as the central community hub for students that attend a school in the urban heart of a mid-sized Canadian city but who do not live in the neighbourhood.

The contextual descriptions of the board and the case schools above provide a backdrop into the ensuing within-case narratives. Next, the within-case findings offer an in-depth portrayal of how teachers perceive their early reading teaching and learning environments.

Within-Case Findings

This section presents themes that represent a compilation of analyses answering two research questions designed to deepen an understanding of how professional learning influences teachers' reading instruction, and the role of context upon their teaching and learning process.

The varying yet related notions in the within-case findings below reflect how much local conditions influence teaching and learning. In this scope, the within-case findings illuminate each of the cases' complexities and articulate varying themes, representing processes involved in developing reading instruction skills and designing a learning environment conducive to improving student reading success. The within-case narrative begins with Outer Elementary School, moves to Blending Elementary and finishes with Central Elementary School.

Outer Elementary School

Outer Elementary School is geographically isolated compared to the other two case schools in this research, and notions of isolation interweave throughout the perspectives of OT1 and OT2, both at the beginning of their careers. The three themes emerging in this within-case analysis highlight the role of community upon the OES teacher participants. These themes reflect how sensitivity to context extends to specific needs in their rural teaching and learning environment and how two new teachers who experience geographical isolation require empathy to their local situation for better addressing their specific professional.

Theme 1. Exclusion: Geographical, Cultural, and Personal-Professional Tension. The first theme relates to perceptions regarding the role of exclusion on opportunities to participate in professional learning. Geographic isolation often hinders OT1's and OT2's opportunities for attending professional learning experiences at the board office. OT2 believes the board considers that OES and other small isolated schools are "on the edge of the earth." Both teachers desire formal and informal opportunities for sharing and socializing; however, OES's physical location and low teacher population impedes access to these kinds of opportunities. For example, OT2 perceives that there was not anything currently in place to bring in new reading resources: "there's no group sharing, there's no leader who is bringing us new material." OT1 spoke about the two years she had spent as the school's literacy representative. Though she appreciated learning new ways to integrate language arts competencies, she also recalls that when trying to share the experiences and resources with her OES colleagues, there was not very much interest.

No teachers share the same grade at OES, however, because schools in the province run on a cycle system (e.g., grades 1 and 2 equate to Cycle 1), this guarantees a degree of partnership between at least two teachers at OES. OP remarks that OES' remoteness limits teachers' opportunities to get professional learning that they need or want: "That's the thing about the small school system. You've got one cycle, you work alone in isolation, same as my position, and that's, that's not healthy. You need to be talking to people." Though OT1 and OT2 both teach in Cycle 1, OT1 yearns for a more substantial community presence and regrets the prevailing sense of isolation and limited collegiality. Though OT2 values opportunities to learn with others, she believes there are limits to teachers' opportunities at schools like OES. OT1 values OT2's presence as her cycle partner because she is an invaluable social and professional asset, and because before her arrival, her perception of isolation was greater. She emphasizes the importance of having a compatible peer and teaching partner: "It's just so important to have someone to talk to you know … it makes a difference having a co-worker that you can work well with." Believing that structured times for meeting with peers from other rural schools would provide a fluid network for sharing that does not currently exist OT1 wishes the province or school board would mandate face to face professional learning for rural teachers. OP suggests OES has to be prepared to utilize webinars to counter the challenges of their isolation. However, both teachers disapprove of this professional learning platform with OT1 remarking "it's so boring you know. They're like, 'pretend you're in a group discussion' … video conference is brutal … It's not the same as collaboration … isolation is so hard."

At OES, most students' home language is French, whereas for most of the other schools across the board, the students' mother tongue is English. The emergent English learning needs of OT1's and OT2's students appear influential upon the desired directions of their professional learning. OP believes that some foundational language learning needs are absent in the board supported reading initiatives. For example, OT1 questions the effectiveness and relatability of current early reading professional learning opportunities for her students and would like more contextualized professional learning: "there needs to be an early literacy for ESL kids, does there need to be one for English natives? Most definitely. It is totally different and it's not fair." In terms of the curriculum, OT1 believes that essential and engaging aspects of reading instruction connected to her students' future needs are neglected. In her opinion, there should be more of a focus on nonfiction to concretely connect more holistically with the lives these students live: "the boys especially who are often more reluctant readers and writers, they love nonfiction."

Within the school board, there is one classroom with an expert early reading teacher for modelling one of the board early reading programs. OT1 has visited the model classroom, and she believes it does not reflect the reality she perceives is necessary for her teaching needs: "it was like watching a show … I want to see the struggle, I want it to be real. Everyone knows when it's a staged performance." In this early phase of her career and alongside her students' specific needs, OT2 perceives one board supported reading program takes too long to figure out, and if she used it, she would have to adapt it to make it useful for her students. Affected by the perceived limits of the current reading initiative OT2 spends hours of out-of-school time learning to integrate contextually relevant reading instruction into her classroom reading program. For example, over the winter break, she created sets of guided reading books to connect with her students' interests, something the collections provided to the schools she perceives are lacking. She spends time after school, searching for online resources and tips to differentiate her early reading instruction. Both teachers exhibit a sense of cultural exclusion in terms of their perceptions that more professional learning and resources relatable to their teaching environment are needed for including their students in more meaningful opportunities for learning how to read.

Representing a further notion of exclusion are perceptions of professional learning being restrictive because of financial and timing constraints. When professional learning conflicts with other priorities, teachers are forced to make difficult choices. For example, OT1 explains that the scheduling of professional learning days is often on holidays, special days, evaluation, and reporting times. She believes that this timing creates barriers to her participation unless she is willing to sacrifice extra hours on weekends and evenings. She also believes that this creates a sense of discord between professional and personal priorities. Financial constraints related to attending professional learning brought up by OT1 and OT2 revolve around obtaining funding to cover their replacement teachers while they are away from the school. Broadly, OT2 believes that the board is not spending enough on professional development; and OT1 remarks that she pays in advance and waits too long for reimbursement after registering for professional development and paying for her lodging and food. To her, this inflexibility restrains her opportunities and her desire to travel to attend professional development: "You can only put up with that for so long before you say I'm not wasting anymore of my personal time or money on professional development no matter how awesome it would be to go … Something needs to be done so that we feel like it's worth it." Emerging from these barriers are perceptions of negativity towards much current professional learning that appears to directly relate to the school's location, the specific needs that their students have, and the ensuing issues of accessibility.

Theme 2. Early Reading Practices Cobbled Together. In the early stages of their careers OT1 and OT2 reflect a developing sense of self-efficacy in relation to how they perceive their reading instruction as meeting their students' needs. Here self-efficacy comprises the participants' beliefs and confidence towards taking and transferring their prior knowledge and professional learning experiences into their early reading practice. In their first year with the board, OT1 and OT2 had to participate in the mandatory new teacher induction program. Both felt limitations concerning their professional learning choices. OP explained that though the induction program directs teacher learning through the lens of twelve teacher competencies, usually there are four specific competencies that most teachers focus on: classroom environment (designing a classroom), structuring teaching (e.g., pacing, delivery), planning (units, lessons), and evaluation (assessment and reporting). Over the short term, OT1 perceives that the program was restrictive because her mentor was chosen for her and because of the limited professional learning opportunities available. She spoke about not having specific help for her early reading instruction and that she had to rely on herself more than learning from her mentor to develop practical knowledge in this area. OP also remarked that the new teacher induction program concludes abruptly following the second year. She perceives that the lack of follow-up appears to leave teachers in a lull for structuring their professional growth: A lull that is beginning to be addressed by OP and other rural principals by using the competencies to help all of their teachers frame their personal professional growth goals.

Reflecting upon her path-as-teacher OT1 acknowledges that "there's still more that I can learn … I've never discovered the perfect resource. I still have to take things from that book or things from this program or something online." Her reading program combines skills and strategies from different experiences as she settles into developing a framework for structuring her early reading instruction. In terms of layering elements of different professional learning experiences, OT1 merges past learning experiences into the present; for example, her prior Smartboard certification helps her use technology to incorporate vocabulary strategy instruction into her teaching. Further, OT1 is adapting early reading professional learning to meet the reading needs of her students. Certain letter sounds not similarly represented between French and English require her flexibility in merging board-provided reading programs to fit her students' language needs. Representing a growing sense of self-efficacy, OT1 is confident in adapting her reading instruction by distinguishing which elements in the board-provided sound and vocabulary program are lacking within her school's context. Incorporating games, reinforcing newly covered materials and linking to her students' early English reading needs are ways she believes technology can help her design fresh and relevant classroom literacy instruction.

OT2 appears to be in a stage of teaching defined by a more immediate reaction to her students' needs. She has taught in two different schools in the board, representing the reality of many beginning teachers, and has never taught the same grade twice. OT2's devotion to her students' needs surfaces throughout all of the interviews, yet her longer-term reading planning and design appear to be reactions to immediate situations: "I don't plan for the year … I don't plan like a crazy big plan ahead of time … No. It's very like that morning." Priorities for providing for students' immediate needs intermingle with self-confidence in how she perceives she is meeting these needs. Speaking about planning phonics-based elements for her reading instruction: "I pull out pieces from there and I rewrite them … So there's a little bit of planning there … I can do that twenty minutes before class … I don't put a lot of focus I just kind of know where I'm going." At this time, aspects of confidence, belief and understanding and willingness reflect the challenges of applying all of what she has observed and uses from the two board supported reading professional learning initiatives: "I've seen her [specialist] do all the lessons and how you do them, so I do use parts of it. There just seems to be so much to do."

OT2 speaks about watching instructional videos and visiting websites to see how educators create relevant resources: "It's sort of like my passion you know … You're like 'What are my hobbies?' 'Looking at YouTube videos on how to teach.'" Her preference for observational learning translates to watching reading experts modelling instruction. Observing others regardless of the platform provides her technical teaching skills: "I like to mimic a lot of the things that I like from seeing other people teach." OT2 currently prefers learning through observation by situating herself in the background and watching other educators modelling reading instruction. She speaks about her guided reading practice being influenced by a range of educators: "teachers … mentor teacher … resource teacher … many people on You Tube." OT1 welcomes personal visits for immediate feedback: "I'm very open, my door's always open, like I want your opinion … I can only learn from that so I really do welcome a lot of input … You know as soon as someone gives me a good idea it's on a post-it and the next time I can get it put together I give it a try."

Additionally, and alluded to in the first theme, the board-supported reading programs and associated professional learning have been met with varying degrees of acceptance by OT1 and OT2. At this early stage in their careers, whole-scale implementation of these initiatives in their classrooms is perceived by them to be unrealistic. Both admit that professional learning experiences centred in these reading programs positively influence some general aspects of their reading instruction. However, they did not address how these programs specifically improve their teaching or their students' reading achievement. Referencing one board literacy initiative, OT1 remarks, "It's not my favorite reading program, it's got some holes, some flaws, but they do teach you some really interesting techniques if you're going to do a read aloud." Though OT2 sprinkles ingredients of this balanced literacy program into her reading instruction, she remains resistant to its wholescale implementation: "I really wish we could do this … but it's … taking the month or two months … or sometimes three months, to really train them [the students] to do it … I feel I'm losing so much time." She perceives that it lacks relevance to her and her students' needs, and she finds it hard to fully buy in: "I feel like it's a bit of a waste of time in the younger grades. I don't think meaningful work is being done … Like it doesn't have to be individual stations … they're supposed to visit all centres, and like it's just I don't know … I like the idea if the kids were actually going to get meaningful work out of it … Yeah [sighing]."

She resigns to using elements of it but does not appear ready to implement it faithfully, nor has she received any formal training on it: "I haven't had any formal training on it … I've seen it used. So I do use some of the concepts from it … I sometimes do it from time to time [whispering]. I don't love it." Though she appears to be a touch cynical and anxious about the board reading initiatives, OT2 attributes the presence of a particular reading specialist to reigniting her passion for teaching reading. Observing SB3's expert teaching has convinced her that the program's elements are effective instructional strategies that she should use in her reading instruction. A writer's workshop led by a board reading consultant and based in poetry motivated OT1 to transfer new elements into her language arts instruction. Finally, both teachers are receptive to having reading experts visit their classrooms to help design their environments.

If SB3 came into model and set up the early literacy stations for one week, OT1 believes that she would feel prepared to execute the program more reliably.

OT1 and OT2 perceive professional learning that provides early reading resources, and that models relevant instructional strategies as influential for transforming their reading practices. Currently, these experiences centre on learning specific aspects of how to provide early reading instruction rather than learning, understanding and implementing board supported reading programs. For example, an opportunity to act as a school literacy representative allowed OT1 to meet with representatives from other schools over two years. This experience introduced her to literacy instruction elements that she likely would not have encountered if she was not a literacy rep. Along with other grade one and two teachers from around the board, the representatives shared ideas and learned techniques to fuse poetry within language arts instruction. This experience inspired OT1 to bring fresh opportunities into her classroom: "new authors, new books, new styles that I'd never seen … and you can see how even though I learned something very simple … in PD how that can really change your instruction." Recounting positive prior experiences with another board early reading initiative, OT2 expresses the benefits of being provided with concrete examples and resources to use in her instruction: "I would be interested in getting my hands on those mini-lessons … Oh I liked it. Just getting into the school and seeing other classes and seeing other examples."

Theme 3. Establishing a Community of Readers. OT1 and OT2 establish their classroom community of readers by empathizing with their students' sociocultural situations and developing engaging classroom reading environments that are foundational to meeting their students' unique learning needs. By creating positive relationships with their students, both teachers perceive that learning how to read is enhanced, and overall satisfaction in their classroom learning communities has improved. As described in the first theme, most OES

students speak French as their first language. This context provides unique challenges for early

reading teachers. Conversations with OT1 and OT2 indicate that establishing a positive

classroom community is crucial to supporting their students' complex learning needs.

Creating this environment has been influenced by past learning experiences, and both

teachers believe that creating environments suitable for their students' unique needs is crucial for

establishing effective reading programs. These professional learning experiences reflect a need to

motivate students, ease their anxiety, and help them settle into their English language reading

environment. OT2 spoke about the importance of connecting with her students at the beginning

of each week. She learned the value of creating connections by volunteering in a classroom that

provided intensive language arts supports, and that began each week, ensuring each student

shared experiences in English about something from their weekend. Recounting her students'

fears and stresses over having to learn in English, OT1 recalled a professor's advice from a

second language additional qualification course. Initially, OT1 had difficulty connecting socially

with her students: "I had my students tell me I was mean cause I asked them to speak English …

I found if you want them to take on the risk of speaking another language and being laughed at

or just making mistakes I need to do the same thing." By putting herself into her students' shoes,

she could connect in a way she could not before. Allowing her students to use their first language

in conjunction with English and challenging herself to use French at times in her teaching helped

unify the class.

For OT1 and OT2, it appears that creating classrooms respectful of their students'

particular language needs is believed to reduce the chance of failure by making learning more

culturally relevant. Cognizant of their students' French language background, both participants recognize the importance of delivering high-quality vocabulary instruction. Most of their students read below average due to their emerging English language skills and socioeconomic background. Developing independent readers based on guiding students from their current levels of need and interest emerged in discussions with OT1: "If I have a group of students that are at kindergarten level I have to teach kindergarten, I can't move forward ... you can't be motivated unless you find whatever is going to click with them and it has to be something that's got a personal significance for them." Setting goals and adapting to changes in routines provides a safe environment for learning, according to OT1. OT2 also perceives that in order for her students to be motivated to learn, she has first to establish positive relationships with them. As trust develops, she sees her students' willingness to improve their reading abilities. OT2 takes time to introduce, highlight and discuss new vocabulary and spend time during a holiday to construct guided reading books based on students' interests and reading levels. OT2 regularly monitors her students' current reading levels and works with individual students when necessary. "I start with letter sounds first ... and then ... words and a writing sample that they can do on their own and obviously a BAS [Benchmark Assessment System] mark so I can start putting them into groups ... then I just do the little things that I think are missing ... as the year goes on I see what they need to work on." Both teachers perceive that generating personal connections with their students and as a class to be an intrinsic catalyst for improving reading achievement.

Stimulating passion and interest are central to OT1's and OT2's reading instruction. Partly this is represented by how both teachers model the pleasure and significance of reading in their lives. OT1 allots time to read silently with her students to model and share how reading is enjoyable. Additionally, she seeks to stoke her students' passion for reading by designing a classroom that accessibly displays an extensive range and quantity of popular children's literature: "I have a display for books … Pete the Cat - I'm obsessed with him and I think I've created three generations of students that are now Pete obsessed because I get so excited … So just finding those special books that make it really interesting." OT1 seeks to motivate her students by fusing reading time with lessons on the importance of reading to provide safe opportunities for sharing their perspectives on what they are reading: "They need to know that reading can be pleasurable … this is important too, just reading for fun … we've had to talk about that as … 'Why do we read?' … you know you've failed if a kid doesn't know why we're learning to read." For OT2, part of her reading program also supports explicit discussion of the importance of reading in students' lives. Because most of her students do not have English as a first language, she ensures class discussions about reading: "We talk about reading a lot. And I talk about the importance of reading and I always say, 'Do you want to drive.' and they're like, 'yes,' all of them, and I said 'well you need to learn how to read so like let's go' [laughing]." Both teachers appear to strive to implement practices within their literacy programs that develop ways for their students to connect to literacy that is pleasurable, safe, and inclusive and are rooted in their students' realities.

Blending Elementary School

Blending Elementary School's location is quite different from the other two case schools. BES serves a suburban population. The two teacher participants have been working together in this school for quite a few years, and the participants, generally, did not relate to any particular needs within a community-minded context such as culture, language, or socioeconomics. The five themes emerging in the within-case analysis of Blending Elementary provide an interesting portrayal of community, reading instruction and professional learning in a mid-sized, suburban elementary school from the perspective of two teachers in the mid-stages of their teaching careers.

Theme 1. Foundations and Change. Tied into numerous past experiences, the notion of enhancing prior beliefs is represented by BT1's and BT2's developing perspectives towards the landscape of early reading instruction and the influential range of their past learning experiences on their early reading practice. This developing outlook reflects their teaching trajectories; in how instruction is guided within the multitudinous backdrop of students' needs, and with dependence upon professional learning to inform their practice and beliefs and improve student reading achievement. Layering knowledge and understanding from diverse professional learning experiences across time shapes BT1's and BT2's perceptions of the diverse teaching knowledge they need for providing an inclusive and effective reading program. BT2 remarks, "a lot of layering since I've started teaching," comprising professional learning on child development, classroom management, social responsibility, designing a whole class reading program, and phonics instruction. Remarking on the learning, change and teaching process, she notes, "it's an ongoing thing … the layering approach," and perceiving that her learning experiences are fitting together "like a puzzle."

Both participants' professional growth develops by having the self-awareness, flexibility and willingness to seek change opportunities. Openness to accessing new tools and contemporary ways of providing reading instruction reflects BT2's commitment to applying her learning in her practice. Using these tools and updated strategies to make reading an engaging learning experience for her students, BT2 remarks that good reading instruction: "is the tools, the practice, the listening and discussing of great books." She has adapted more systematic instruction into her whole language belief system that underlies her reading instruction approach.

For approximately her first ten years of teaching, she did not value code-based elements of reading instruction. It was not until experiencing its value through the board's code-based (word-study) reading initiative that she decided it was time to include it. She perceives that her most recent approach blends a rich reading environment with layers of "other different strategies or tools." Over time BT2 has committed to enhancing her foundations of reading instruction by integrating child development, attachment theory and organizational strategies and by working to integrate these professional learning experiences into her early reading environment and practices. For example, she remarked that "organizational strategies are to me a key to making the day successful … it just eliminates a lot of conflict, which is not good for teaching, which is not good for learning and it gives a sense of security and we know where we're going, what we're supposed to do."

BT1 also underwent a shift in her teaching beliefs and changed her instruction to include more explicit code-based reading instruction. Speaking to this shift, BT1 recounts: "we sit down and we do actual phonetic instruction … It's real explicit and that's changed a lot from the beginning of my career where I was like, 'Oh they'll get it through osmosis.'" BT1's ongoing search for effective ways to help struggling readers illustrates the influence of a commitment to guiding her learning and change concerning all of her students' success: "You just keep trying things until you find something that kind of works … then it stops working and then you have to find something else." For BT1, professional learning was instrumental at the beginning of her career: "I didn't know how to do a guided reading lesson when I started teaching. No one told me how to do it in teacher's college. It was through PD and learning communities that I learned how to do that stuff." BT2 similarly expresses this sentiment. Not until she joined a particular professional learning network focused on the two board reading initiatives did she understand the effect that small group guided reading could have on her students' reading development: "[i]t's been huge [the] last three years." A strong element to this theme reflects an understanding of how these teachers' varied early reading experiences influence their perspectives of their instruction and how a range of experiences are appreciated as they were seeking to become better early reading teachers in their beginning years.

Six years ago, BT2 returned part-time to teaching after an extended parental leave. She attributes her participation in an early reading professional learning community to helping her design an effective and efficient classroom reading program since her return. Previously, she remarks, she would take thirty minutes to introduce a lesson, but now was providing more effective focus lessons in 10 minutes: "my first kind of experience seeing these little focus lessons … was a bit of a mindset change … that was an eye-opener for me." Professional learning explicitly influences areas of BT1's early reading practices as well. Initially, she believed students would, over time, become efficient decoders. Participating in a board supported reading initiative focusing on elements of phonemic awareness and phonics contributed to shifting this belief. BT1 believes she has become better at providing explicit instruction in teaching letter sound and recognition strategies, and that has helped her students who struggle to learn how to read. Generally speaking, BT1 and BT2's involvement in two early reading board initiatives over the last decade has helped them transform their practices. BT1 believes she has learned the importance of synthesizing and structuring oral language, writing, and reading within small group instruction. BT2 perceives she is developing a more comprehensive perspective that connects the dots for her by merging her whole language beliefs alongside an appreciation of the benefits of explicit reading instruction. The perspectives of BT1 and BT2 in this theme indicate the role of professional learning on their early reading instruction.

Expressed in fairly positive perspectives, the role of professional learning on early reading learning trajectories and instruction reflects ways that professional learning alters teachers' perceptions of the processes and components of reading instruction across time.

Theme 2. Intrapersonal Responsibility. Accountability, obligations and perceptions of divided responsibilities are tensions BT1 and BT2 express in discussions about obstacles to meeting their professional learning needs. Along with a sense of refined learning responsibilities that develop as teaching experience and maturity accumulates is the notion of increasing personal responsibilities and the challenges that they present to participating in the professional learning that they would like to.

Alluded to above is a bumpiness resulting from trying to balance personal and professional responsibilities. BT1 and BT2 are mothers of young children. Balancing their personal responsibilities with their sense of professional obligation generates friction and often leads to limited involvement in professional learning for long periods. BT2 is not motivated to attend workshops that require her to be away from home. Further, following several years of involvement in ongoing early reading professional learning, she believes she has progressively developed an effective literacy program; but that "there's still more fine-tuning to do" and, therefore, doesn't want to "jump on too many ships at once." With her collected experience over time, BT2 values focused and concentrated learning opportunities. However, connected to this degree of appreciation for the more in-depth learning experiences are the barriers to her professional learning constrained by her current life situation. Unless notified about something interesting, she is satisfied with her current state of learning: "If I have things on my plate already that I feel I could continue to improve why would I go and take another chunk of stuff." BT1 wishes that available professional learning experiences are more considerate for mothers and teachers' family obligations. She believes that much of the professional learning that occurs after school is not accessible for many teachers; that too many teachers face obligations outside of the school, which often makes it difficult for teachers in the same school to connect after students leave for the day. BT1 revealed this was the first year she had "done the afterschool stuff [in many years] because my mom happened to move to the community so can help me out and pick my kids up from daycare."

For both teacher participants, professional learning experiences within their school and particularly in their classrooms are the ones that are most appealing to them. For BT2 having an expert visit after school limits the pressures that come from financing a supply teacher, the disruptions that come by leaving her students for a day or two, and the pressures of being away from her family. BT1 would like to use her prep time to explore reading practices and philosophy. She perceives that it is a rare space where she might be able to work around her personal and professional constraints. Ultimately though, she would like the consideration of teachers' learning needs to be more inclusive. She believes that professional learning in the school and during the day would positively address many of the accessibility issues and benefit a large population of teachers. Still, she questions the likelihood of this occurring bringing up the obligations she has during her teaching hours: "I would really love to have the time in my day … to use that [prep] to sit down and read a couple chapters … but I'm calling parents, emailing parents, cleaning up the mess that the last class made, trying to catch up on my math marking." Family and professional obligations are always at play and appear a burden upon teachers' sense of professional and personal responsibilities and barriers to participating in sustained and deep levels of professional learning. These barriers emerge over time as teachers have families and as they grapple with obligations to their students and their parents each school day.

Theme 3. Valued Relationships: Peers and Experts. Throughout the discussions with BT1 and BT2, both express that their relationships with other professionals influences their reading knowledge and beliefs. These shared perspectives of BT1 and BT2 represent the value they place on their learning community because of: (a) its diverse range of individuals within the social collective who embody varying roles and, (b) how new knowledge and early reading practices emerge. Learning a common language and recognizing roles and supports within their community through sustained professional relationships with their peers, mentors, and board-level experts is perceived by both teachers to be positively influencing their early reading instruction.

According to BT1 and BT2, one-day professional learning workshops located outside of their school board do not provide the sustained social connections they perceive are necessary to transform their instruction. For both of them, membership in a network of early reading teachers from BES and other schools in the board provides a universal language that nurture and sustain professional learning relationships. BT1 and BT2 perceive that these types of social learning experiences improve their reading instruction because the combination of active elements to their learning is effective. Listening to others, hearing about new practices, observing how others teach and dialoguing with colleagues and experts provide rich and intimate ways for BT1 and BT2 to learn: "Just having that inspiration of other teachers … having that around you rather than nobody else trying it or nobody to balance ideas off of" (BT2). For BT1, the side conversations emerging in small learning communities are very influential to her guided reading knowledge and practice. Sustained opportunities for collaborative learning ensures that multiple voices are learning in shared spaces. Sustained learning experiences within her school and throughout the board have been influential to her teaching by providing a broad range of ideas to "modify" depending on her class and student needs. She credits ongoing reading professional learning initiatives for helping her connect the dots to early reading instruction and for creating valued learning communities: "for a few years, every couple of months, usually with people within your board or your school … building a community of likeminded people so you're not working in your school in isolation."

For BT1 and BT2, their professional relationships are powerful vehicles for teacher and student change. At the school level, both participants express how a shared language around reading instruction emerged between teachers. It appears to represent an organic form of learning possible because of a collective sense of understanding generated. BT2 remarks how open the conversations are at BES because everyone is coming from a similar base of understanding: "we're a lucky school … we're all talking the same language … we've been in these workshops together … we're all implementing them in some form in our classroom … it really gives us a spring board … our kids in grade one they go to the next class … [and] they're doing it there and the next one." This shared approach for learning opens up ongoing dialogue and BT2 attributes this for her "plugging into the reading in a bigger way." BT1 noted that her cycle team holds impromptu meetings for sharing ideas on specific areas of early reading and make lists together that they try in their classes and discuss later. Speaking with a colleague in the hall or emailing the lead literacy consultant is an efficient way to get answers and resources. Learning with others across time appears to instill a sense of belongingness and community. However, BTI perceives that consistent collaborative professional learning "even once a month you know would be fabulous" is difficult to sustain. Nevertheless, professional learning with peers from other schools and over time is valued and would "certainly [be] essential if you didn't have like-minded people in your own school" (BT2).

Speaking about the power of an effective mentoring presence on her teaching journey BT1 remarks: "[W]hat has really helped me to become a better teacher, and I like to think I am a fairly decent teacher … it's mentoring that is the most helpful." SB3, a board reading consultant who visits schools and works with teachers in their classrooms, is responsible for promoting a "mindset change" for BT2 in framing her literacy classes. BT2 would rather turn to the resource teacher at BES or expert consultants from the board for specific information to help a struggling student rather than sign up for workshops. For her, these experts are usually available and provide valuable resources directly connected to her early reading professional learning and her students' reading needs. BT1 emphasizes the influence of SB3 and another early reading consultant and mentor (M1) early in her career in shaping her early reading instruction: (a) "When SB3 comes in and shows you how to do it in your classroom … that was really for me the most meaningful," and (b) "M1 … came in and she was 'this is actually what you should be doing,' showed us how to do it and I was like 'great now I know what to do.'" Claiming, "my first year teaching I didn't teach anyone to read, I had no idea what I was doing," she believes that she learned the fundamentals for providing small group intensive reading instruction from M1's expert guidance in her second year teaching. In fact, BT1's early mentorship experiences reverberate almost 20 years later an interest in herself becoming a mentor to beginning teachers. She perceives that this kind of leadership experience would be an effective way to sharpen her early reading perspectives and help to make herself "a little more cautious of what [she's] doing." The opportunity to mentor is one way that she believes will offer her a sustainable learning experience and help her continue "to be in the know."

Theme 4. Responding to Student Need. According to BP, current professional learning trends are moving towards making schooling inclusive for more students. She remarks that much professional learning has shifted to a focus on understanding the learner (e.g., "anxiety, or autism … transgender … exam conditions"). This shift is noticeable in BT2's recent interest in understanding the nature of child attachment and development and how she perceives it influences her approach to early reading instruction. BT1 spoke of the board recently directing some focus towards teaching communication (conversation/expression) skills to help students learn how to express their emotions and communicate their specific learning needs effectively.

Expressing how vast and ever-changing the notion of the term *struggling* is, BT2 says: "you could have ten struggling readers and they're all struggling in a different way." BT1 notes that a significant challenge to teaching reading is finding consistent instructional reading strategies and programs for struggling readers. What works with one student often does not translate fluently to others. There is a sense of heaviness or burden that both teachers emit as they express the annual challenges trying to meet the unique needs of their struggling early readers. Bringing up a workshop on child development she has attended several times, BT2 mentioned the varying conditions students in grade 1 have. The range of students' brain development varies greatly. She believes this contributes to much of the unsettled and impulsive behaviours occurring in the more traditional confines of a grade 1 curriculum compared to their prior play-based learning environments in daycare and kindergarten. Sequentially, she believes this makes learning to read a challenging exercise for too many of her students. This belief influences how BT2 considers how to structure and organize reading instruction within environments designed to be fun, relaxing, secure and safe: "some of those kids come in grade 1 and other parts of them still need time to develop … only having the benchmarks and the instruction without the other is not happening. I don't think you're getting the best out of your kids if you do that … you need both sides." BT1 believes that her motivation for differentiating

reading instruction and developing passionate readers is central for learning how to read:

> I think if I'm so enthusiastic about it … it rubs off on them. If I look like I don't care
>
> they're not going to care … and it becomes a very personal relationship right … Like
>
> you get very close to someone when you are teaching them how to read … You become
>
> really emotionally attached to them as readers and they become very emotionally
>
> attached to you as the person who is helping them read.

Trusting relationships spark motivation according to both participants, and this generates

strengthened relationships with and enjoyment towards reading: "You need time to know them,

time to have some fun with them, and to relax with them and not just be geared towards an

academic goal … [W]hen you know the kids feel secure and safe … and I can see the kids

relaxing … you can move forward" (BT2).

Though BT1 and BT2 believe that early reading professional learning helps them with

their overall early reading instruction, they believe that board-provided professional learning

opportunities for struggling readers are lacking. Remarking that board reading initiatives help to

provide a broad range of strategies, BT1 believes that "we're not becoming trained in specific

learning disabilities or difficulties." Both participants desire more in-depth professional learning

and perceive that they have a lot more to learn to help their struggling students. BP espouses a

contrasting viewpoint on the nature of struggling readers. She believes that most struggling

readers overcome their challenges by junior high school. Consulting with teachers when

considering if a student might have a learning disability she admits that some kids aren't mature

or "there's a reading readiness that's not there." Taking a broader perspective on the nature of

early literacy learning, BT2 spoke about the relationship between learning to read and reading readiness. Reading readiness reflects being in a regulated state, conducive to effective interaction and participation in one's learning. A presentation on child development and attachment influenced BT2's developing appreciation of the role of social and emotional factors at play in her students' learning. The developmental information was not too new to her, but relating developmental research findings within an attachment framework "helped validate or explain what was going on with kids and helps me have compassion for the kids as they are in class." Repeated visits to this presenter cemented her belief that you cannot separate your awareness of students' social and developmental stages and needs without considering how they work together to influence students' academic expectations and achievement.

Theme 5. Issues of Accessibility and Uptake. At BES, discussions regarding the accessibility of teacher learning opportunities reveal a sense of friction between levels of administration and teachers and a sense that access to learning often appears out of teachers' control. Reminiscing on returning to this school board some years back, BT1 says she received her learning opportunities in early reading instruction because she asked for it but believes that "if you didn't ask they wouldn't be giving it to you." On certain provincial professional learning days during the year, board personnel and principals attend and invite different teachers to join, but BT1 and BT2 speak to their limited opportunities to be a part of this. BT1 remarking on independently searching for professional learning: "you need to go, you need to seek out the collaboration, you need to ask for the help, you need to go see the people [otherwise] you could sit with no support."

BP's perspective as the school's principal provides another sense to the flow teachers navigate when accessing professional learning. Believing that there are few limitations and many supports for teachers, she sees "it more of like an internal thing, like teachers themselves who would be the roadblock not necessarily the system." She spoke about the different committees and networks her teachers are a part of and that accessing board initiatives occurs by teachers reaching out to her and at times when the board asks her to suggest teachers for varying initiatives. Related to this, and appearing to underpin her instructional leadership, are perceptions on what student success is and how to prioritize students' early reading needs. For example, each year, teacher teams meet to project which students will perform well or will struggle, and it appears that BP as the principal makes the final decisions on which groups will receive additional interventions. BP sees some students as having inherent reading difficulties but also believes that most students who are struggling in reading will settle into being proficient readers as they mature. This likely plays a part in decisions about who will receive intervention, and it could account for the professional learning goals BP is working on with her staff. Currently, she wants to broaden hers and the staff's understanding and practical knowledge in helping students develop the metacognitive skills she believes will help them take ownership of their learning.

BT1 believes there has been an overall drop in the availability of early reading professional learning workshops at the board level and in general. A regular participant at annual workshops put on by a company called the Bureau of Educational Research, BT1 remarks that she has not heard of them visiting the area for several years. Currently, a perceived sudden reduction of prior early reading initiatives that were consistent and sustained across time appears to be jarring and disruptive to the fluidity of these participants' ongoing professional learning journeys. For example, BT1's general impression on the sustainability of different learning initiatives provided by the board: "It's never consistent, no, it needs to be." BT2 spoke more specifically: "the board tends to, I think, get highly focused on one area … unfortunately some things get picked up … then they get dropped and the next interesting thing gets picked up and dropped, picked up and dropped." What appears to happen is that as the board drops favoured initiatives, a perceived lack of transparency transpires, and confusion, as well as negativity, is projected from teachers toward the board.

One experience that stands out for BT2 occurred a few years earlier when the board concluded an initiative related to instructional, organizational and classroom management strategy development. It ended abruptly, and she wishes that there had been more memorable in-house work to keep her connected to those prior professional learning experiences. There is a growing sense the current decline will lead to the end of the early reading networks. One early reading collaborative, in particular, appears to be meeting less consistently, and visits from one expert early reading consultant have ceased. The effect that this has on within school team meetings also appears negative as BT1 remarks that their "cycle meetings happen once in a blue moon."

Finally, BT1 and BT2 attribute some blame for these disruptions to economic concerns. BT2 believes that operating costs are a big reason current sustained early reading professional learning is fading. BT1 relates how teachers shoulder the burden of financial pressures when registering for professional learning. For example, paying for professional learning external to what the board provides is first covered by the individual teacher. Reimbursement comes later. When there are bills to pay and Christmas gifts maxing out the credit card, the obligation to put personal needs first often appears to override down payments on professional learning.

Central Elementary School

Representing Central Elementary School are two teacher participants with a combined 99% of their professional teaching time in this highly populated inner urban elementary school. Emerging from the analysis are four themes relating the influence of contextual uniqueness interacting upon CT1's and CT2's early reading teaching and learning. Thrown into the reality of what early reading teaching appears to comprise and their accompanying unpreparedness to provide it, the value of undertaking a range of early reading learning opportunities, the kinds of transformational influences (in particular mentor relationships), the value in understanding students' particular social-emotional needs, and the fractures and barriers to obtaining professional learning interact to influence how CT1 and CT2 perceive their early reading learning and practice. Together these themes portray the essence of teaching and learning within CES' large urban, low-socioeconomic context.

Theme 1. Trial by Fire: Frenzy, Foundations and Commitment to Harnessing Momentum. Perceptions of being overwhelmed with the number of skills and understanding necessary for teaching early reading are associated with the foundational knowledge that early reading professional learning experiences provided CT1 and CT2. The importance of having a range of early reading professional learning experiences emerges from understanding the relationships between professional learning, teachers' early reading instruction, and their students' achievement. For CT1 and CT2, their early years are motivators for their change and growth into early reading teachers. Entry into the completely new learning environment that CT1 and CT2 confronted as beginning teachers prompted them to seek learning experiences that would improve their reading instruction and boost their self-efficacy. Professionally, the difference in years of teaching between CT1 and CT2 is nine, and yet their initial experiences upon entry into early reading teaching are noticeably similar: the perceived overwhelming nature of early reading professional teaching and learning. Looking back, CT2 perceives a lack of refinement in her first professional learning choices. She would often be in audiences of hundreds of other teachers viewing PowerPoints. It took her several years to realize how important it was to determine a sustained focus that reflected her classroom teaching needs. CT1 also spoke about how she would attend several large conferences each year in her first five years, going to as many workshops as possible. CT2, reflecting upon her professional readiness for teaching reading in her first year, remarks, "I wasn't a good teacher cause I didn't know what I was doing … [I was] not prepared at all." CT1 comparably recounts that she: "spent those first two years figuring out how to teach reading." Abrupt, essential, and forced learning experiences appear to be catalysts for change in this early stage of teaching.

The cognitive dissonance arising from their transitions into teaching pushed CT1 and CT2 towards many early reading professional learning experiences. Confronted with the responsibility of teaching the fundamental skill of early reading, at first, the range of suggestions, ways to teach reading and the variety of resources to use in their reading programs were dizzying. CT2 provides a sense of the dizzying nature of these first years of her early reading professional learning journey: "I wasn't a good teacher cause I didn't know what I was doing and it was fresh … You get so many resources and you get overwhelmed in resources … and it's like you have to stop and just focus on one or two things not try to use everything." CT1 presents a parallel picture of the unsettling reality of professional learning in teachers' early years. Mentioning that beginning teachers are "just trying to stay afloat," CT1 spoke about how the early years are about figuring out how to implement elements of a reading program. Looking back on her early years, she shared her perspective on how critical it is to learn how to set up the framework for early reading and to provide a structured reading environment for students to learn to read. The sense of urgency to learn how to provide reading instruction (structure, skills and knowledge) reverberated with these teachers during the beginning years of their teaching as they confronted the complexity and challenges that comprise early reading instruction.

Though overwhelming, both teachers attribute immersion into a range of workshops as experiences responsible for instilling their foundational understanding of the essential elements for providing effective early reading instruction. For CT1, this involve seeking out as much early reading professional learning as she could in her beginning years, learning about the value of different parts of a reading program, and the significance of setting up a program that provides her time to work with her students. She credits one professional learning group, the Bureau of Education and Research (BER), for providing her with resources and strategies to help her in her early years: "Their presenters were so good … they gave you materials and such incredible teaching stuff … that's where I learned the bulk of my early literacy skills and experiences." CT2 also attributes her learning to her school board and the BER in helping her understand the basics of early reading instruction and how to structure her program.

Workshops provided essential skills that helped CT1 and CT2 begin to implement necessary structures into their reading programs. Learning how to teach decoding, word work skills, shared reading, guided reading, and structure a balanced reading program were a few of the skills they learned and used in creating their instructional environments. CT2 perceives that she benefited greatly from two board-supported early reading learning initiatives in terms of structuring language arts and obtaining practical resources to facilitate her instruction. CT1 recounts the influence of early reading professional learning through a focus on structure, provision of resources and the variety of specific elements: "I went to them all … Whenever one would come up I would go to it and learn and learn and apply it in my classroom … I became quite focused on guided reading and becoming a good guided reading instructor and as soon as that happened you could really make your students progress."

CT1's early experiences and dedication to providing and improving upon her guided reading instruction has been ongoing for close to 20 years. CT2, with fewer years of experience, has taught grade one for 8 of her 11 years. Seeking new strategies or adapting pre-existing ones into their reading instruction emerges as new needs arise. Professional learning's role appears to be essential in these teachers' growth and their students' success. For CT2, this shows in how she adapts reading strategies discovered in professional learning to meet the diverse needs of her students:

> When you become more experienced you realize what's working for your class and what you need to do to help that cohort of students in the classroom so you start picking workshops or PLs [professional learnings] according to what the needs are … PD reinforces what you already know and allows you to assess the efficiency of learning strategies and then it makes it more relevant to students.

CT2, through her experiences with a school board reading consultant and her role in board reading initiatives, is now one of the board's model teachers for early reading instruction and literacy centres. She believes it took her four years to implement her reading program effectively and confidently. CT2 will still spend weeks to months (depending on her class' capacity for independence) at the beginning of each year, committed to introducing one element of the program at a time. CT1 believes that her early reading professional learning experiences across her 21 year career are directly responsible for improving her students' reading achievement: "It's taught me better ways to teach them so then they've achieved higher levels directly relative to the professional development that I've taken." For CT1, it is her commitment to implementing a guided reading program that leads her to perceive success in providing small group instruction while her other students work independently: "The whole mystery and challenge was … being able to do that … Honestly when I've had classes that have struggled … I've had the most success moving them through small group instruction like that."

Confronting the complexities of early reading professional learning occurred early in both teachers' entry into the teaching profession. Both perceive they were underprepared to teach early reading at the beginning of their careers and focused their professional learning upon its components and effective ways of providing it. These experiences ignited a passion for teaching early reading and were essential to their current practices. Both perceive that developing teaching practices for improving student outcomes depends on consistent participation in ongoing and related professional learning opportunities. The combination of experience blended with CT1's commitment to learning the intricacies of a program over time is integral to her students' reading achievement: "I feel like the PD, the training, the experience, I really feel like I've gotten better with age. Like I feel more confident in what I'm doing … I think the onus is going to be on me to, through professional learning really, like through the internet, through reading, whatever I can attend." CT2 initially introduced to one board reading initiative for designing a cohesive literacy structure in her first year of teaching grade one did not, until her fourth year of implementing it, feel capable of providing all components at once. At first, engulfed within a complexity of responsibilities, she believes that her commitment to learning and implementing this reading program provided her with the structure she required to support her students' early reading needs. CT1's drive to continue to strive and to "crack the code" or CT2's commitment to learn, experiment, and after four years, confidently implement all elements in her reading program reflects their passion for staying connected to their past learning experiences and their belief in the importance of seeking opportunities to extend their understanding of the reading process throughout their careers.

Theme 2. Engaged Mentor Relationships. Connecting to the role of professional learning throughout their teaching careers is the influence of mentors on both teachers. Though CT1's and CT2's mentoring experiences are markedly different, it does appear that their mentors have positively impacted these teachers' professional learning experiences, early reading instruction and, subsequently, their students' reading achievement. CT2 finds the relationship she has established with SB3, a former teacher and current reading consultant with the school board, a very fortunate influence upon her teaching. CT1 appreciates past involvement in small learning networks and is unequivocal about the power one memorable weeklong mentorship experience has had on her teaching career.

CT2 attributes her relationship with SB3 as central to her access to a range of early reading professional learning groups and her role as a classroom model teacher for the two board reading early reading initiatives. Until she had the support from SB3, CT2 believes she missed participating in different learning groups, having a mentoring relationship, and knowing how to access resources for professional learning. Appearing as an essential aspect of this relationship is that much of this professional learning appears to have happened in CT2's classroom. For CT2, resources to use for practicing and creating mini-lessons were helpful. The addition of a reading expert in leading lessons with conversations considering these lessons helped her see how things worked or why the expert provided particular aspects in their teaching. Learning with resources provided and an expert to model has also helped CT2 believe that she does not always need to "recreate the wheel."

Having someone to learn with and from has been transformative for CT1. Following an extremely rough stretch of teaching in her second year, she attributes a turnaround in her attitude and practice to her principal and a retired teacher her principal brought in to spend five straight and "non-threatening days observing and helping [her] in the classroom." After school every day

for a couple of hours, they would discuss short- and long-term planning and chart ideas on paper.

The mentor helped CT1 use her strengths and to understand the areas where she was struggling.

She claims it changed her by helping her understand and develop crucial skills at this critical

juncture early in her teaching career. CT1 was positively affected by this one-week mentoring

experience early in her career and suggests that there should be more opportunities for teachers

who perceive they are struggling with a particular part of their teaching. Ongoing access to

experts is something that CT1 would appreciate having more remarking that "in a perfect world

every second Friday I would go to school and someone would teach me how to do things better

in my classroom … it would be more practice than reactive." She believes that this would allow

her to focus on classroom needs, and if it involved colleagues, this would be extra motivating.

Learning by watching those you professionally admire is one form of professional learning she

would like more opportunities to have. She also understands that though this transformed her

teaching, it could be considered an obtrusion to other teachers.

Although different, both teachers perceive the positive impact more experienced teachers

have had on their early reading instruction by helping them structure, organize and implement

instruction that meets their students' early reading needs.

Theme 3. Student Well-being and Societal Shift. This theme points to the perceived

importance that the three CES participants place on recognizing that a large proportion of their

students arrive at school in emotional states that unfortunately set them up for negative

academic, social and behavioural consequences; and that improving these circumstances appears

to rely on a school-based approach. CT1, CT2 and CP believe that one of their primary

responsibilities is to create positive learning environments that motivates all students to learn.

126

The complex needs within and among students at CES are powerful influencers upon CT1's and CT2's teaching practices, their ability to connect with students, and their professional learning. They both recognize that each student brings diverse collections of experiences to school and that these experiences are responsible for influencing emotions, physical needs and, therefore, attitudes towards learning, peers and teachers.

CT2 spoke about the uniqueness of student cohorts each year. She relates how these yearly particularities mediate how sufficient her current instructional knowledge is. For her, this limits how predictable professional learning will be over time. Whereas in her early years CT2 would attend all the professional learning she could access, she now seeks opportunities associated with her current students' learning needs. For example, her most recent professional learning included integrating yoga and brain breaks into her literacy instruction. CP's interest in children's social, emotional and developmental attitudes appears to anchor CES's focus on helping students regulate and positively attach to school. He finds that compared to a generation before, many students are arriving at school unready to learn:

> Why's the teachability changed? I'd say factors in society have changed, the business of parents have changed, the amount of care that our kids are in, away from the parents and grandparents, or aunts and uncles has changed. So context has changed … Kids are not coming into school at the same emotional relational place that we assume they are… (CP)

On top of this societal shift is the bumpiness many students experience in their transition from their kindergarten, play-based environments, to the more traditional landscape of academic-accountability and expectation within the grade one curriculum. Expectations of learning in this grade one environment appear to reveal that too many students are not ready for this and have other needs that require attention before smoothening the shift into the grade one curriculum.

CT1 shared how she perceives her failures in meeting the struggling early reading needs of particular students and that these perceived failures provoked her to incorporate play-based principles within her early reading instruction because of the many students she felt were not ready for learning to read in the more traditional setting.

CP believes that the board approaches professional learning from the perspective of what is most important to them in terms of overall academic successes. He related that the current system prioritizes standardized performances, compares students to averages and that this notion of success limits the opportunities for the emergence of improved social and emotional skills too many students need. CP also believes that many students ("in our school anyway for about 40% of the kids") need teachers to do the work that used to be done at home. At CES, CP is explicit that societal shifts and changes to the family structure connect to a deceleration of many students' social-emotional development, which subsequently impacts reading readiness. He believes that the curricular expectations for students in grade one are above their current capacities and that asking for independence from students contradicts what many need: "dependency on the safety of the relationship, a dependency on the trust and expectations and the structure." CP notes, it is a crippling paradox when expecting children to be teachable through their transition into the more traditional curriculum and associated pedagogy. For CP, this paradox reveals some limits to current expectations for these students, within a current societal context, many whom he believes are not emotionally ready for learning in more traditional school structures.

CT1 and CT2 also relate their perceptions that an increasingly large population of students at CES lack sufficient confidence, independence and motivation to learn how to read. These elements restrict the formation of positive relationships between teachers and students;

and consequentially chances for learning in school. CT2's own experiences of being an outsider and learning a new language after moving to Canada when she was eight helped shape her perspective that factors originating outside the school heavily influence student achievement. CT2 believes that students' academic struggles will be reduced by teachers who understand how to design inclusive classrooms and school communities: "you think about your struggles … I don't want them to go through that … If the kids don't have a positive attitude … they'll shut down." Sustained learning experiences focusing on child development and attachment alongside her principal helps her bring social-emotional components into her teaching. She believes these are effectively helping her students connect to school, learning, and their academic development, and she believes it helps her students develop a positive outlook on life and in school: "I've done a lot of reading … about Neufeld and attachment theory and that's helped because forming the attachment with the kid and feeling safe … they start learning." CT1 echoes this belief in comments about how generating positive classroom environments and relationships are integral to her students' success:

> If they're not at school at rest ready to learn then there's nothing you can do … [if] they don't come to school with their basic needs met then I don't think there's anything you can do except work at that attachment, work at getting them to trust you and then when that happens you can move them ahead.

Both teachers agree that student success requires students to have self-confidence and self-belief, and they perceive that their roles as teachers extend beyond the reading curriculum to help students become successful readers. CP sees the challenging expectations that grade one teachers have teaching groups of emergent academic learners who will struggle in many facets of the classroom. He suggests a change of approach is required to help alleviate the pressures on students. To him, a healthy system is one where teachers would not have to fail students, where students would not confront failure from the beginning and where parents would not have anxiety because their children are floundering within a system in which their primary needs are not being met and are in turn inhibiting chances for present and future well-being. The current contextual environment for many students at CES demonstrates that teachers and the administration perceive that professional learning for early reading has to broaden and encompass social-emotional aspects.

Theme 4. Fractured access: Freedom, funds and time. The participants' perceptions at CES reveal how priorities reflecting from the student, school and board levels influence access to teachers' professional learning. For example, CP thinks that the overall direction for professional learning comes from the board. The teachers believe that the board pays attention to academic success, on which CT1 remarks: "the specific outcomes or successes of the students forces people to, forces us to learn in or to improve certain areas," and that at the school level the principal's interests drive current learning initiatives. CT1, speaking about a prior principal who "was very into developing professional learning," recalled weekly literacy instruction and assessment meetings. CT1 appreciates the directive and direction of CP's social and emotional professional development initiative, yet also indicates how other teachers feel forced to participate. CT1 believes that professional learning:

> has to be driven by the teachers, by the people participating in it. If it's shoved down your throat by your principal because they want you to be learning that I don't think it's nearly as successful as if you're doing what people are interested in or what they can feel in your classroom.

Speaking to the notion of availability and choice, CT2 wishes that she had more freedom to choose workshops "according to what your class clientele is." Even when something is available, the choice reflects priorities at a provincial or board level that too often do not reflect her immediate needs. CT2 would like to suggest workshops and resources that would help her to meet specific classroom needs. CT1 suggests ongoing professional learning that provides follow-up on more specialized topics (e.g., dyslexia) and would allow her to develop a greater understanding of reading difficulties stepwise across time.

Access to funding and time are key factors that participants perceive are influencing the nature of and opportunity to engage with professional learning experiences. For example, CT1 appreciates the one thousand dollars for professional learning teachers receive for helping to finance professional learning every two years; however, both teachers find that their professional learning funds quickly dry up because of the costs for covering supply teachers and registration fees. According to CT2, there are additional opportunities to generate funding through a professional innovation grant. Still, CT1 did not discuss having accessed these funds, and CT2 remarked that she does not fully understand the application process and is reliant on others to coordinate a plan. Regardless of their perspectives on available choices and presumed needs, there is a comparable sense of their vulnerability to the quantity of readily available professional learning. In comparison to earlier in her career, CT1 remarked upon the reduced availability of professional learning over the past ten years, expressing that she has not experienced any memorable professional learning for over a year. Similarly, CT2 shared her confusion about how the early reading professional learning network had gone from meeting five times the year before to none in the current year.

Issues of inclusion and sustainability appear to influence teachers' perspectives of their learning opportunities. CT2 remarks that many workshops she attends are funded through board grants and that her most recent involvement was an invitation by the board for herself and five other teachers. She suggests this is an exclusionary process for the many teachers not chosen. Related is a sense of expectation from the board to be a part of the board-supported reading initiatives. Prioritization of two early reading professional learning initiatives likely limits other beneficial learning opportunities. Further, those who do not choose to receive specific training cannot access specific reading resources to use in their classroom. Whereas CT2 expresses enthusiasm towards both initiatives, CT1 is more reluctant, considering her interests are elsewhere at this time in her career. For example, CT1 emphasized that she does not understand how to adapt her teaching to meet her students' most diverse early reading needs. She would prefer information on teaching students with ADHD, how to interpret results from psychoeducational testing, or understanding the nature of reading disabilities.

Summary of Within-case Analyses

Before presenting the next phase of findings, Table 4 below lays out the themes generated in the within-case analysis.

Table 4

Summary of Within-case Themes

Theme	OES	BES	CES
1	Exclusion: Geographical, Cultural, and Personal Professional	Foundations of Change	Trial by Fire: Foundations & Commitment to Harnessing Momentum
2	Early Reading Practices Cobbled Together	Intrapersonal Responsibility	Engaged Mentor Relationships

Theme	OES	BES	CES
3	Establishing a Community of Learners	Valued Relationships: Peers and Experts	Student Well-being and Societal Shift
4		Responding to Student Need	Fractured Access: Freedom, Funds and Time
5		Issues of Accessibility and Uptake	

These themes represent the teacher participants' stories in their schools and express their connections to past and present experiences through their perspectives on their early reading professional learning and practice in their particular time and space. These stories tell how context influences their professional learning experiences, how it appears they have learned to teach early reading and how this influences their perceptions of meeting their students' early reading needs. The within-case narratives are deeply personal. Teaching is deeply personal, and moving from with-case findings to shared lived experiences is informed by displaying the contextual pathways influencing perspectives of effective early reading teaching and learning. The next section lays out the findings from the transition from within-case to cross-case themes in the form of contextual networks.

Emergences: A Transition towards the Cross-case

Described in the analysis section and Appendix U, this transitional step incorporates Miles and Huberman's causal network (Miles et al., 2014), adapting its stylistic visual flow of pathways. Whereas Miles and Huberman create causal networks to explain within-case qualitative cause and effect relationships, my adaptation is a contextual network that visually portrays how contextual variables are perceived influential to participants' professional learning experiences, early reading instruction and student reading outcomes. Two central reasons for creating these visual contextual displays are to have some clarity that might not be transparent in the unique case narratives, and second, the pathways to and from succinct and somewhat generalized terms help to situate the audience into a location for seeing patterns and abstractions across cases. There are no cut and dry causal explanations here, but rather a range of pathways reflecting how different processes and experiences appear to influence teachers in these case schools. Further, no weights are represented in these relationships, yet understanding the influence of specific contextual factors perceived by participants to be influential to their professional learning, practice, and students' reading success follows a linear directional flow represented by pathways of arrows.

In terms of organization of the flow of my research, the process of creating the networks draws heavily from the within-case themes, is inclusive to perspectives presented by all participants, and allows me to (re)calibrate my analyses to represent the dynamic nature of the interactions between collections of variables. Illustrating the jumping-off point from the within-case narratives towards contextual networks (which are the transitional links to cross-case analysis), the themes subsume from across the range of variables for each case school. This process involved mining the within-case narratives and crafting labels of contextual variables across one or all themes (see Figures 3, 4, and 5 below).

Figure 3

OES Diffusion of Variables and Factors across Themes

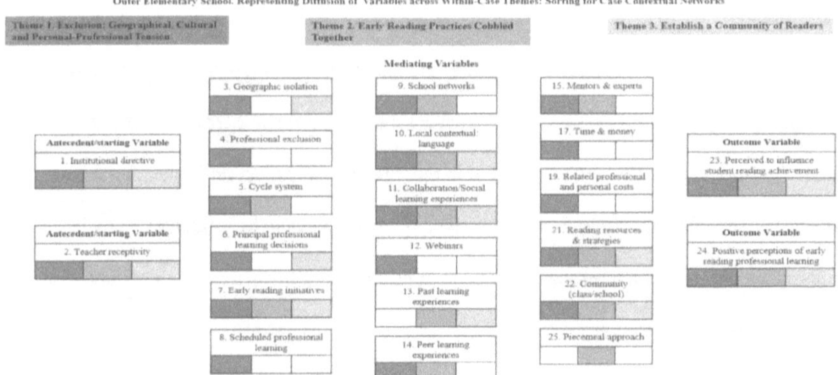

Figure 4

BES Diffusion of Variables and Factors across Themes

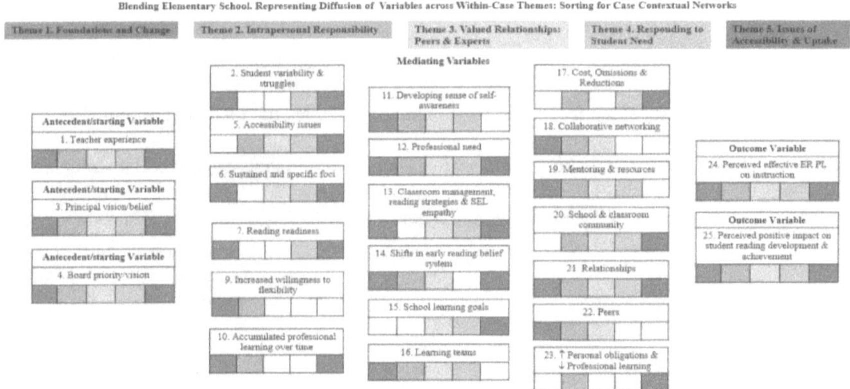

Figure 5

CES Diffusion of Variables and Factors across Themes

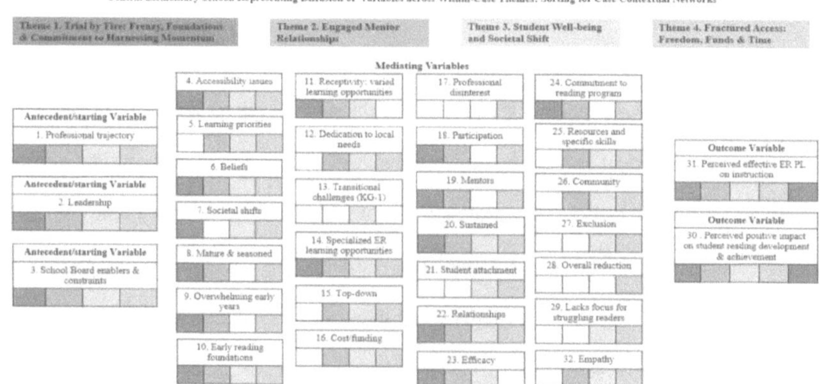

Each theme is colour-coded and many contextual variables represent elements in multiple themes, though how they are perceived may differ. The antecedent variables and outcomes comprise each theme because at minimum they represent elements influential upon and in professional learning and reading achievement.

Contextual narratives in Appendices V, X and Y, explain each pathway and are informative, enriching and highly recommended supplemental reading. These contextual networks portray how context is a collective activity and how then these collective activities or processes appear to influence and follow pathways that help to bring together the elements of the research questions: First, how teachers perceive contextual variables at the level of the school, board and beyond influence the planning, delivery and uptake of early reading professional learning; and second, the influence of their reading instruction on their students' early reading success (achievement, outcomes).

136

Before the presentation of the short descriptions portraying the main ideas for each contextual network, I would again like to point out that there are narratives of each pathway for each school in Appendix V, W and X. They are not included here because of my consideration of the length of this book and how this might tax the reader's concentration and synthesis of thought. However, I strongly encourage a reading of the short narratives for each path. Specifically, the contextual network narratives show three possibilities: Two I consider affirming possibilities to effective early reading professional learning and contributing to improving students' early reading success, and the other more restrictive (negative) to teachers' learning, teaching and students' reading development. In all three network narratives, you will find more explicitly how each theme subsumes from each within-case analysis. In the two kinds of paths I consider affirming, these pathways represent flow through variables that in combination are, (a) perceived to lead to improved early reading instruction from professional learning, or (b) perceived to accumulate in a way that there is a positive impact on student reading development (and achievement). Both of these outcomes reflect in the pathways beginning with variables on the left-hand side of the contextual network. For example, see the pathways in Figure 6. Follow the path (1<5<11<14<21<24<23), ending on the right-hand side of the page. The restricted pathways represent the other side of this picture. At some point, before the two positive outcomes (positive perceptions of professional learning and positive perceptions of having an impact on student reading development), these streams end and do not result in positively perceived influences on teacher learning or student reading development. For example, again, look at Figure 6 below and follow paths 1<3<5<8 and 1<3<10<17<19. Both of these end without reaching outcomes representing positive perceptions of learning or culminating in a contextual pathway representative of factors perceived to be influential in influencing students' early

137

reading development. These contextual networks are an accumulation of subsumption and a process of exhuming the within-case themes and data representing the cases into pathways that allow abstractions across the cases to begin materializing.

Proceeding the visual display of each school, a short synthesis of direction, flow and interaction is provided. Each visual display represents directionality influenced in a particular context. This type of display also produces patterns of shared abstractions, reflecting participants' perspectives. These abstractions generated in these displays anchor the cross-case findings. The direction of this stage of within-case findings begins with Outer Elementary, moves to Blending Elementary and finishes with Central Elementary School.

Outer Elementary School

OES' rurality is a powerful influence upon how teacher participants at OES spoke about their teaching and learning experiences and Figure 6 below clearly represents this. Variables within pathways represented in Figure 6 help separate some of the local contextual variables and those that appear more related to influences from the system at the board level (board initiatives, style of professional learning). The structure of the contextual networks is bound in showing the direction towards answering the research questions. Though local, cultural and geographical factors are intertwined and interactive, the value placed on collaborative professional learning designs that include mentorship opportunities, are authentic and provide instruction on teaching struggling learners is highly visible (see Figure 6 below). For example, in Figure 6 look, at box 11 (Collaboration/Social Learning Opportunities) and box 14 (Peer Learning experiences). It is clear that social learning experiences are perceived to be important but also highly relate to specific local aspects. Figure 6 maps out the complexity of context at the local level showing how students' language ability is central to teachers' professional learning and teaching and how

138

meeting this need is critical within other important aspects designed for collaborative learning
between peers in schools across the board. For a more detailed description of each of the
pathways, see Appendix V.

Figure 6

OES Contextual Network

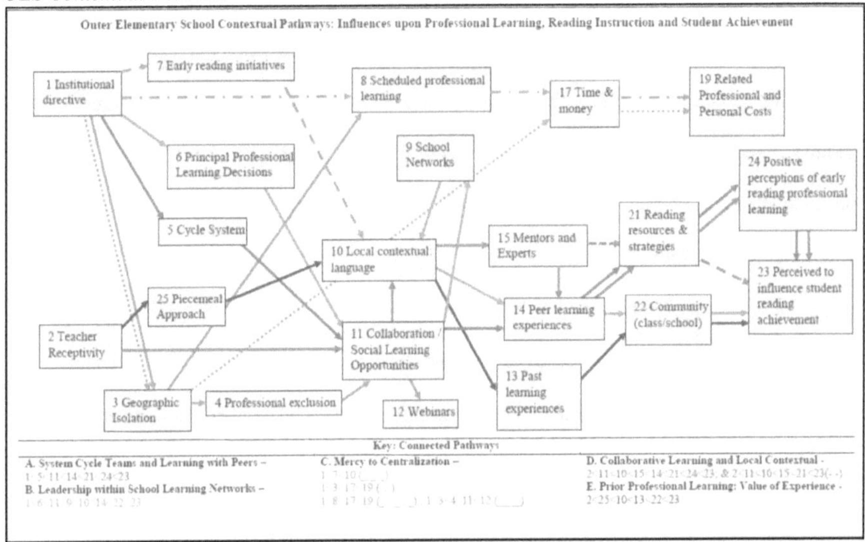

Blending Elementary School

Blending Elementary school teacher participants are mid-career teachers and have been
teaching at BES for quite a few years consecutively. Their perspectives provided a sense of their
learning needs and their attitudes towards how they are met over time. Time-oriented aspects of
learning are reflected in how BT1 and BT2 consider their philosophies on teaching early reading
have expanded due to their accumulation of particular teaching and learning experiences. The
value of learning communities upon teaching and learning is emphasized as a component of

139

professional learning perceived as effective and sustaining and is represented by specific

contextual variables in Figure 7 (boxes – 18, 19, 20, 21, and 22). School and classroom

community is central in BT1's and BT2's perceptions towards meeting their students' early

reading needs. Moreover, interacting with designs and topics of professional learning, when

professional learning happens, and the degree of control for choosing what these teachers

perceive they need, are pathways reflecting from the principal (school) and board levels (see

Figure 7 below).

Figure 7

BES Contextual Network

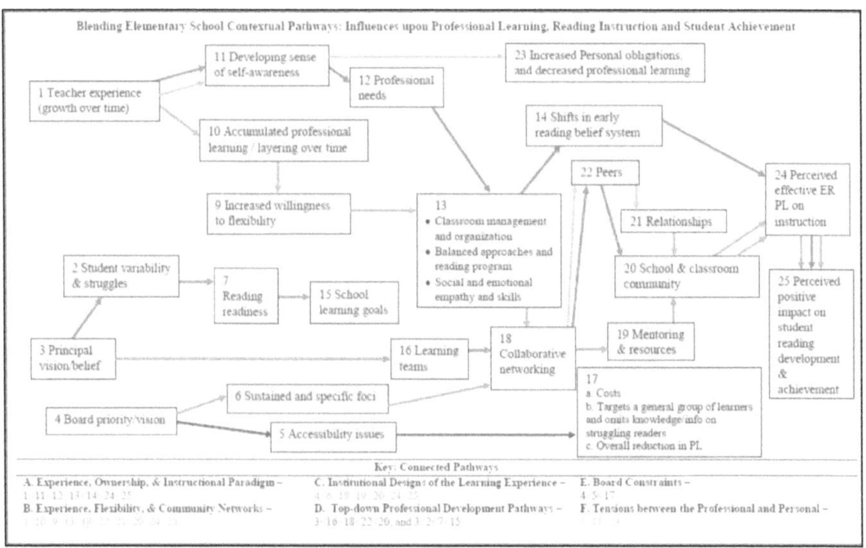

The richness of this case study comes out in the depth of description on perspectives

across time. It provides a uniquely complex, but specific contextual network displaying how

these two early reading teachers have developed into the experienced teachers they are now and

140

who need new learning opportunities to continue improving. See Appendix W for the more detailed description of each connected pathway.

Central Elementary School

CT1 and CT2 have spent nearly their entire careers at Central Elementary School. Figure 8 below represents the connected contextual networks derived from their perspectives.

Figure 8

CES Contextual Network

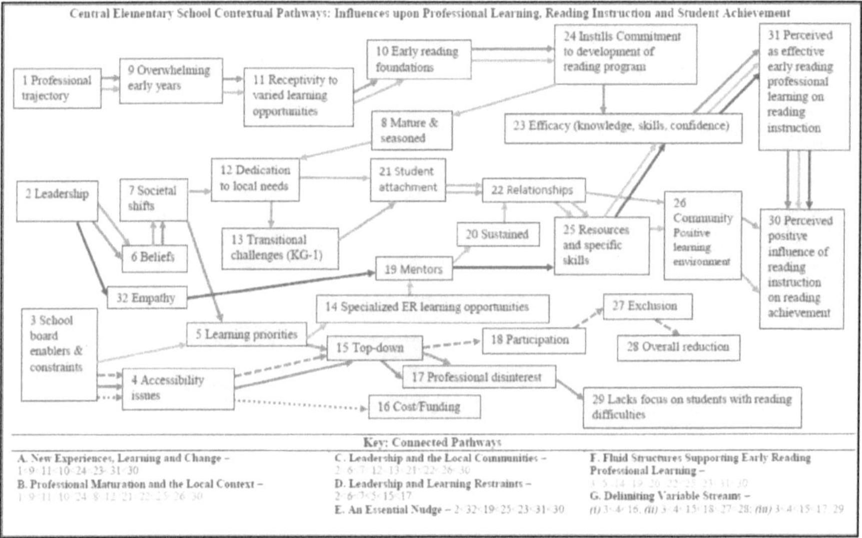

These teachers' perspectives relate to how across the trajectories of their careers, they have developed as reading teachers, learning to understand the particular needs of their students, committing to learning perceived-to-be effective reading programs, and immersing themselves into a range of professional learning early in their careers. The principal- and school board-provided professional learning experiences heavily influences these teachers' current reading instruction. Mentorship opportunities provided both teachers with the foundations for their professional development and practice. What is interesting with this case school is how commitment to provide an effective program persists, continuing to drive their teaching and perceived-expertise (e.g., boxes 10, 24, 23).

Students' particular contextual needs at CES underpin CP's social and emotional ideals on teaching (e.g. Boxes, 6, 7, 12). The pathways presented in Figure 8 break this down. Contextual influences emanate from self, leadership and board levels, flowing and interacting with the needs of this urban, low SES school. The stages of development of early reading teachers in this environment and the provided social structures to learning are integral to certain aspects of learning to teach reading. Box 26 in Figure 8 above emphasizes how the community needs of CES are a final stepping point in teachers' perceptions of how they consider their professional learning to influence their knowledge of teaching reading and how effective their teaching is upon their students' early reading needs (see pathways to and from Box 26). Importantly one pathway also reflects the limits to how professional learning has not provided enough ways to help them meet the needs of their students who have specific difficulties learning to read (Box, 29). See Appendix X for a more detailed description of each connected pathway.

The next section, the cross-case findings, brings the abstractions of cases together and represents the three diverse cases as representative of schools across the school board. In considering the board level, the cross-case findings demonstrates a further depth to thinking of professional learning and its influence on teachers' early reading professional learning and practice. The cross-case analysis and findings emerge from three within-case contextual networks, which derive from the within-case narratives. Whereas the within-case findings for each case school provide a narrative that is unique and deeply entrenched upon local context/s, the cross-case findings reveal abstractions and more general patterns reflecting the natures of the participants' perspectives on their learning and how they perceive that certain experiences appear to influence their instruction and consequently their students' reading outcomes. The within-case contextual networks above visually represent a refined, skeletal look at the streams (or pathways) that are richly elaborated upon in their own ways within the individual within-case narratives. Instead of relaying the themes, they utilize the interacting relationships influencing participants' perceptions of early reading instruction, early reading professional learning and the relatedness of these experiences to student achievement in reading; and in this next section are expressed more broadly as the cross-case findings.

Cross-Case Findings: Three Central Abstractions

Emerging in the cross-case analysis are several key themes that centre on three factors broadly influencing teacher participants' perceptions and perspectives of their professional learning: the teacher, the student and the learning community. *Emergence, notions of initial chaos and characteristics of experience* is the first theme. The developmental nature of becoming and being teacher is a variable representing a non-linear accumulation of learning experiences shaping the bumpy and temporally linear developmental process teachers follow through interactions that lead to a particular wealth of early reading knowledges and skills. The centrepiece of the cross-case findings, the second theme, is *Professional collaborative connections.* Learning with other professionals (peers, colleagues, and mentor experts) is an experience all teacher participants value and utilize in varying forms from different levels in the system. Collaboration appears as a catalyst within teachers' learning experiences that lead to in-depth knowledge, a developing sense of community, and in providing the kinds of teacher learning experiences that improve practice, and ultimately students' reading outcomes.

Inclusive learning environments is the third cross-case theme. This theme reflects why learning and understanding how to teach within one's particular context is essential. The specific challenges that students in the case schools are perceived to face and how uniqueness influences the professional learning direction that teachers take encapsulates the notions of this theme. As the third theme, *Inclusive learning environments* represent the diversity of academic and social, emotional and cultural learning needs and how understanding a school's local context influences the ways teachers perceive their professional learning. Broadly and very importantly, this theme also comprises the consensus across teacher participants that they need more learning experiences to help them develop effective instruction to help their most struggling readers. Unfortunately, an aspect of this theme suggests that even with perceived-to-be effective structures in place such as collaborative learning experiences and opportunities to learn how to set up reading programs, these structures are not enough for them to help students with reading difficulties and disabilities. A description of each theme occurs below. Following the cross-case findings is the presentation of the conceptual framework, the culmination of this research study's in-depth stages of data analyses and the writing up of the findings.

Theme 1. Emergence, Notions of Initial Chaos and Characteristics of Experience

Participants in this research study embody professional teachers with experiences ranging between two to over 20 years of teaching in diversely situated elementary schools. CT1, BT1 and BT2 have the most experience, each teaching approximately 20 years and most of it at their current schools. At the time of the interview, CT2 was in her 11th year, all at CES; OT1 was in her third year of full-time teaching at OES (4th overall as a professional teacher), and OT2 was in her second full year of teaching in the school board (in her first full year at OES). The range of these participants provides unique perspectives of participants' relationships to their specific learning experiences and depicts an understanding of the trajectory of their development as professional teachers concerning their early reading instruction. These perspectives illuminate how the course of being a teacher is never stable and that with varying stages arise challenges that represent a need for considering teacher learning and practice as non-linear and difficult to predict but very important for considering when planning how to better meet teacher learning needs and evolving states of students' early reading needs. The responsibility and challenge of becoming a confident early reading teacher are not unique to any of the participants, and the sense of discomfort appears as a motif across all six participants' professional teaching and learning journeys.

OT1's and OT2's current perspectives on the nature of their early reading professional learning represent their belief that they should be participating in and implementing their school board's early reading initiatives. However, what appears to influence their current instruction is less systematic. Their current practices are reflective of learning via a piecemeal approach in the ways they borrow from a range of learning experiences and their acknowledgement of how much they still have to learn. At this early stage of their careers, both perceive that the whole-scale implementation of a particular board supported reading program is currently unrealistic. Partly this is because OES participants perceive the available board reading initiatives lack some core elements that their students require. Nevertheless, comparing their approaches to early reading instruction at this early stage with the other teacher participants, what emerges is how complex, scattered and overwhelming the beginning early stages of teaching and learning are.

CT1 spent her early years figuring out how to teach and took everything she could, "trying to stay afloat." Reflecting on the hectic learning and teaching experiences early in her career, she attributes the range of early reading professional learning to shaping her effectiveness as an early reading teacher and directly relative to her students' reading success. CT2's early years of teaching were overwhelming due to the range of suggestions, the number of workshops, and the challenge of incorporating many strategies into her reading instruction. She values the range of experiences, even though it took her a few years to realize she needed to regulate her professional learning choices. In her third year of teaching and first in grade 1, she was fortunate to connect with a reading consultant, and has slowly developed her expertise in setting up what she perceives is a reasonably effective reading program. Eight years later, CT2 continues to regulate the speed of implementation of literacy stations as this is dependent upon the composition and related needs of her current class. For both CES teachers, they attribute their immersion into a range of early reading workshops on guided reading and setting up a balanced approach using structured reading stations as foundational to their early reading instruction.

BT1 and BT2 both attribute specific professional learning to improving their early reading instruction. BT1 began her teaching career confused about how to provide effective reading classes. She explicitly references her commitment to board reading programs for helping her connect "all those little dots," as well as the range of experiences offered by the board for helping her layer these learning experiences into her teaching where she now provides a more systematic approach than she did early in her career. Moreover, for BT1 learning is an ongoing process in which she constantly searches for strategies to help her struggling readers. Her flexibility takes root in the awareness that the needs of her students change across time. She perceives that learning how to implement guided reading early in her career, as well as ongoing experiences learning about code-based instruction, as integral to her practice and her students' success. Originally supposing that learning to read comes naturally through exposure to good literature over time, it took her several years to value the importance of bringing different code-based components into her instruction. BT2 spoke about her commitment to providing an enriching whole language environment in her approach to reading instruction, but how opportunities that she had not previously considered now enhance her instruction. She perceives that through the range of her professional learning experiences, she continues to improve her teaching and students' experiences and success. Learning experiences on child development, classroom and organizational management, social responsibility, and board reading initiatives are layers of experience that she believes contribute to her providing successful early reading instruction.

It is apparent that the early stages of teaching are essential times for grounding oneself professionally, but for this to happen there needs to be a range of opportunities available for beginning teachers. OT1 and OT2 appear to be entering the profession when there has been a reduction in the quality and quantity of professional learning compared to the past. Whereas CT1, CT2, BT1 and BT2 speak to the abundance of reading-related professional learning early in their careers they all note that there has been a substantial reduction in the availability of professional learning over the past several years; and that only particular learning experiences have been available in terms of specific elements of the board supported reading programs. OT1 and OT2 have experienced aspects of the two board supported initiatives yet project that they need more to develop their confidence in providing all their students with practical learning experiences. All teachers expressed that learning to provide reading instruction is an ongoing endeavour. The four most experienced teachers in this study describe the reduction of professional learning as confusing and difficult. Whether it is a loss of networks they are accustomed to or an overall reduction in the variety of early reading learning opportunities, they perceive the current state of professional learning is limiting to their development.

CT1, BT1 and BT2 have had children throughout their teaching careers. There is a perception that much of the professional learning that occurs after school is not accessible for many teachers; that too many teachers face obligations outside of the school, and this often makes it difficult for teachers in the same school to connect after the bell signals the end of the school day. It appears that in terms of learning and practice it is important to consider how the responsibilities and obligations that come with having a family can affect the trajectory of teachers' professional development and innovation to their practice. Currently, there is a sense of discord expressed and perceived by teachers when having to make choices between professional learning opportunities and family; if there is an opportunity to choose at all. The three teacher participants who are mothers echo a sense of being cut off or barred from professional learning. CT1 related being left out of the professional learning circles while raising her two children. Only in the past few years does she find she can devote some more time to joining organized professional learning. At BES, BT1 is just beginning to think about attending afterschool learning opportunities at the board because her mother has moved back to Blending and takes care of her children some days. BT2 remarks that she is happy with the pace of her job and that she is not trying too hard to find new learning opportunities because she has a young child.

Emergence as a professional is an ongoing state of being, development and becoming. It appears that particular stages of teaching are needing more significant consideration to help anchor effective early reading practices. The dizzying pace beginning teachers experience, as represented by the participants in this research study, suggests that (a) beginning career teachers will benefit from understanding what the transition into being a professional teacher is like and that (b) receiving a range of early reading professional learning helps anchor core early reading practices. As professional learning opportunities appear to be diminishing, less experienced

148

teachers OT1 and OT2, who are in the early stages of their careers, appear to be taking what they can to establish some core reading instruction practices. As professional teachers such as CT1, CT2, BT1, and BT2 mature, they want to build upon their core early reading practices. Recognition of the challenges that surface when balancing family and professional responsibilities, and having a way to support these challenges, will expand their practice, contribute to meeting their students' needs and provide a sense of being included in and controlling their professional development.

Theme 2. Professional Collaborative Connections

Presenting this theme in the middle of my cross-case findings flows from a conscious research and educational decision to represent professional collaboration as the centrepiece to learning as a professional. There are mediating factors that present differences in the nature of collaborative learning across cases, however, learning with others (another) is an undeniably valuable component of these participants' learning processes. This theme portrays how collaboration is viewed across cases and is streamlined within a range of variables that influence the nature and availability of collaborative learning possibilities.

Each new teacher to the board, regardless of experience, partners with a mentor teacher as part of a two-year teacher induction program. Participants hold mixed perspectives about this program. For example, BT1 considers that her mentor teacher was missing the particular skills that she needed to develop her early reading instruction. OT1 perceives that her first two years of professional development limited her growth as an early reading teacher because of what she had to focus on in her mentoring program and not on any specific early reading professional development. Still, it appears that this mentoring program offers new teachers some opportunities to receive guidance through relationships with more experienced teachers that the

149

board deems qualified to mentor. However, none of the teachers perceive the program's influence as overly positive to their development as early reading teachers.

All teachers do mention their appreciation for experiences with expert early reading mentors for helping them understand and implement instructional strategies and practical resources. Having mentors who are early reading experts as a part of a learning initiative is deemed highly valuable to their learning. CT2, OT1, OT2, BT1 and BT2 described their favourable experiences with mentor teachers in the board. CT1 does not mention a particular early reading mentor though she attributes one mentor early in her career for transforming her teaching perspective and practice. Mainly, mentors who visit schools and classrooms provide the resources and skills they believe are effective in reaching their students' particular needs. OT1 and OT2 have mixed reviews of the two recent board initiatives, but they both stress how much they value their mentoring from the consultant leading the initiatives. BT1, BT2 and CT2 attribute the presence of this consultant for providing them a more transparent and relevant sense of how to incorporate aspects of the board-supported reading initiatives into their teaching and design of their early reading programs. BT1 and BT2 also find that sustained professional learning involving a mentor expert in early reading help shape a positive community at BES for teachers to converse, share and reflect on their early reading experiences.

Underscoring the value of collaborative learning are the experiences that OT1 and OT2 have had at OES. Their geographical isolation provides a backdrop for viewing the perceived significance of accessing collaborative learning experiences. OES's physical isolation makes it challenging to take part in face-to-face social learning experiences. The population of this rural school denotes the challenges in developing effective partnerships within and outside the school. Current technologies for connecting with peers who can access face-to-face learning in central locations such as the school board do not appear to provide the types of collaborative learning experiences both teachers value. Besides, the small teacher population does not provide stimulating collaborative learning experiences across the school. OP, the principal of OES, is aware of the restrictions to collaborative learning, and she is trying to administer school-based professional learning where teachers are observing each other, reflecting and sharing on their growth and goals in more structured afterschool staff meetings. However, in the teacher interviews with OT1 and OT2, this topic is not addressed.

Provincially, the ministry of education structures grades into cycles with grades paired consecutively (as a cycle) beginning in grade one. For example, grades 1 and 2 represent the first cycle. The cycle system ensures that at least two teachers will be within one cycle at most schools, and therefore teachers will have opportunities to at minimum dialogue about the directions their programs and students are moving in. BT1 and BT2 highly favour cycle teams and it appears that along with the presence of a mentor to move them along their knowledge, understanding and teaching improves because of the opportunities to collaborate with their grade and cycle colleagues. The benefits of cycle teams were not overtly mentioned by CT1 and CT2, though CT2 spoke about the importance of cohesion in early reading instruction at Central Elementary School. She admits some pessimism towards some colleagues' lack of buy-in to working together. It is possible that because there are many teachers within each cycle at CES, this could make it harder to organize and maintain cycle meetings or reduces the chance for the professionally intimate conversations like the ones that appear to happen more frequently at BES. Diametrical to CES is OES's small population of students and teachers. Peer collaboration is valued, and though OT1 is very appreciative of having OT2 as a cycle partner, both teachers project a yearning for increased collaborative opportunities with colleagues and peers teaching the same grade.

Across cases, it appears that when there is a mentor who is an expert in early reading and who visits schools, there is a profound sense of appreciation for how this connects to their local contexts and, in the case of BES, helps to stimulate reciprocal peer learning experiences amongst cycle teams. It appears that collaborative learning is how teachers want to learn and that an institutional nudge to structure this support is needed. Due to their isolation at OES, OT1 and OT2 express a need for assistance from a mentor. Though collaborative opportunities are available to them through online conferencing, these have not yet been positive experiences. At Blending Elementary, BP prefers that the teachers, for the most part, work in their cycle or grade teams to enhance their professional development and will often let the teachers self-direct their learning. BT1 and BT2 do not openly discuss these networks in terms of specific school-based professional learning. What they perceive to be stimulating and helpful to their early reading instruction is a mentor who is an expert and works with them and their peers to help create a shared language they find necessary to their development, and that works towards creating a cohesive school approach to reading instruction. At CES, CP desires to create learning networks with teachers mentoring each other, but a lack of stability in sustaining positive group learning and regular teacher changeover challenges the reality of making this happen. CT1 attributes one particular experience with a mentor teacher to keeping her in the profession and helping her structure her overall approach to teaching, and CT2 attributes her eight-year ongoing relationship with a reading expert to developing her early reading teaching skills.

Across all of the participants, it is apparent that collaborative professional learning is a preferred mode for learning, collaboration is instrumental to building community at the

152

classroom, school and board levels, and that through various learning networks the participants in this study find collaborative learning the necessary median for supporting their early reading learning needs.

Theme 3. Inclusive Learning Environments

Inclusive education is anchored in and relates to aspects of interactions influencing how participants in this study perceive their students' needs connect to their learning and practice and, subsequently, their students' reading development. The concept of inclusive learning environments presented here moves from a leadership-teacher-board direction in considering the role of meeting students' particular needs by relating to students' readiness for reading success. To meet students' early reading needs, teachers have to be experts in providing reading instruction that supports the needs of struggling early readers.

Students' needs appear highly dependent upon the context they learn and relate to the places in which they grow up and live. Viewed differently depending on population differences and demographics or cultural backgrounds, the particular uniqueness of CES and OES especially appear to propel school-based professional learning approaches. Recognition of students' particular needs is quickly evident with OT1 and OT2 at OES and CT1 and CT2 at CES, noting how they are regularly working to finding ways to connect to these needs and how their principals organize professional learning to help their school communities embrace their students' particular needs. At Outer Elementary, OT1 and OT2 find the board early reading initiatives underwhelming with Principal OP recognizing that they presume a level of readiness beyond their students' emergent English language needs. Therefore OP collaborates with teachers to determine what programs they can learn together. At the time of our interviews, one of their school growth goals focused on developing student vocabulary skills and word banks.

In terms of school populations, CES is one of the largest elementary schools in the board. Feeding into their particular urban location are students with transitory parents and guardians from single-family homes and low-income earning households, including a considerable population of First Nations students from northern communities whose caregiver/s cycle/s between staying for short periods and returning home. The school focuses on developing and integrating teacher understanding and practices to develop social-emotional relationship skills. In his seven years as CES' principal, CP has heavily invested in creating a community-minded school inclusive to his students' needs. He believes the school will best achieve this by creating reciprocal relationships between students, teachers and parents. His perspective is deliberately influencing the direction of professional learning at the school level. Though CT1 remarks that some teachers appear jaded or disinterested towards the professional learning CP offers, by focusing on developing attachment and relationships, she finds it to be a crucial part of her teaching and learning. Both CT1 and CT2 side with their principal's belief and CT2 is even looking to integrate aspects of social-emotional learning and regulation into her literacy instruction.

The influence of particular localized needs of students was not as prevalent in the interviews with BES participants. Principal BP, for the most part, prefers that cycle and grade teams develop on an as-needed basis and the trending issues they are noticing. BP inserts degrees of initiatives such as universal design for learning and works with teachers to understand evidence-based approaches to teaching (e.g., John Hattie's effect size work). However, the influence of some specific contextual needs at BES did not appear to stand out for any of the three participants in this school, especially in comparison to the influence of community and cultural context at CES and OES. This opens up questions and reflects how particular directions

154

(or lack of directive) can influence the learning opportunities teachers will have and subsequently, how this influences their instruction and will impact student learning opportunities and reading achievement. However, across cases, all teachers shared their collective concerns about the lack of professional learning targeting intervention for students with specific reading difficulties.

All teachers mention that there just was not enough professional learning available to help them teach and improve their struggling students' reading outcomes. These struggling readers have a horrible time learning to read, more of an internal difficulty that relates to learning to read. In this situation, participants presume that they are missing something in their teaching repertoire, and professional learning is not providing the answer. At OES, teachers perceive that the early reading initiatives are not very effective for many of their students who need core English vocabulary development. OT1 and OT2 also believe that there has not been any specific training for them to help students with reading disabilities and difficulties. CT1, CT2, BT1, and BT2 also perceive that professional learning experiences neglect their struggling students' needs. Learning to better provide for the needs of students who have dyslexia or have other intense reading difficulties do not appear to be a part of any past professional learning. All participants express that they would welcome opportunities for learning more about reading difficulties. All teachers project a sense of guilt that they cannot provide effective instruction for their students with specific learning difficulties and disabilities. Teacher participants at OES and CES perceive that their schools' focus on their particular unique needs helps build community. However, these teachers, along with BT1 and BT2, believe they are not adequately meeting all of their students' needs in ways to include them in their learning more fully; that doing a better job of this would connect to an increased community of learning and inclusively attach more students to their schooling experiences.

Emerging from within- and cross-case findings, a conceptual framework brings together the relationships between teachers (their learning and change) and students (their success and well-being). This framework considers the teacher and learning system as one that is at its core a system of human interactions, combinations of reactions within and across variables, and influenced through the interactions of contextual factors at play upon the individuals comprising the system.

A Conceptual Framework Contextualizing Teaching, Learning and Student Well-being

In outlining the process of data analysis in Chapter 4, the reasoning behind the multiple in-depth stages of analysis in this research is, ultimately, to create spaces for inductive analysis to guide the researcher toward findings in which concept and hypotheses emerge out of the teacher participants' stories of their lived experiences (Attinasi, Jr., 1991). This research concerns how teachers perceive their early reading instruction relating to professional learning and student reading success. It is also partly the product of the data from the three case school principals and three board-level participants who provide contextual information from semi-structured interviews and supplement the teacher participant data. Coming before a description of how the conceptual framework operates is a short section outlining the process leading up to its conception. Miles et al. (2015) express how a conceptual framework graphically explains "the main things to be studied – the key factors, variables, or constructs – and the presumed interrelationships among them … and are the current version of the territory being investigated" (p. 20). The conceptual framework that evolves in this research visually depicts the relationships between contextual factors interacting upon the main variables that influence learning to read, professional learning and teachers' change processes.

This framework is the product of an iterative research process seeking to contextually conceptualize the process of early reading professional learning, teacher practices and student early reading development through a complexity theory lens for opening up and unpacking the system/s of early reading instruction within a province, school board, schools and classrooms. Construction of the framework began with a comprehensive review of the literature on early reading instruction, teacher early reading knowledge, professional learning, and teacher change to display how too many students continue to not learn to read. Emerging from an initial review of the literature are two open-ended research questions for inquiring into perceptions of the influence of contextual variables upon teachers' early reading learning experiences and the role of these experiences upon their teaching and their student' reading development. In turn, these research questions guide the creation of the in-depth three-series interviews with six teacher participants in three case schools in one school board in Québec, proceeds through multiple stages of data reduction and analysis (profiles, category construction, coding, ongoing memoing and iterative narratives), leads to within case findings, visual displays of data representing within-case pathways of perspectives influencing professional learning and student reading development and finally, comprises the collective abstractions that emerge in the cross-case findings (see Stages of Data Analysis, Figure 2, p 64). Following multiple, iterative drafts of the cross-case findings, the research process returns to the overarching problem, within a complexity theory mindset, to depict why reading failures prevail in schools for too many children and the associated negative societal implications.

This mindset comprises a point of view focusing on the levels of variables interacting with teachers and how contextual factors influence teachers' learning processes within early reading. The variables, teacher state and stage, student well-being, and the learning community depict: how teacher learning and instruction emerge partly in teachers' histories and current circumstance; the influence of student well-being (academic and social-emotional) upon the teacher and learning process; the diverse ways a learning community influences teachers' professional learning and reading instruction; and how the structure of the learning system situates itself for reacting to students' early reading needs. Further, comprising the elements in the social system are individuals (or selves) comprising ties within and across each of the variables in this framework.

Figure 9 illustrates the interconnection of the within- and cross-case themes and contextual network labels and relates how varying modes of analysis and layering relate, comprising the process's totality in this version of the contextual framework. This figure provides an audit trail leading toward the conceptual framework, substantiating each step of analysis and corresponding findings. Consequently, these factors are not random or lightly thought out. They are notions that connect in varying degrees to the variables and interconnect with all elements influencing the system. A conceptual framework is a significant structure in theory building (Corbin & Strauss, 1990; Miles et al., 2014). In the process of inductive analysis, theory does not drive the design. Instead, the design is driven by systematic attempts to generate theory. In this research, the data analysis representing the perspectives of the six teacher participants' lived experiences underpins the structure and interactivity of the learning, change, and achievement conceptual framework.

Figure 9

Substantiating the Data for the Conceptual Framework

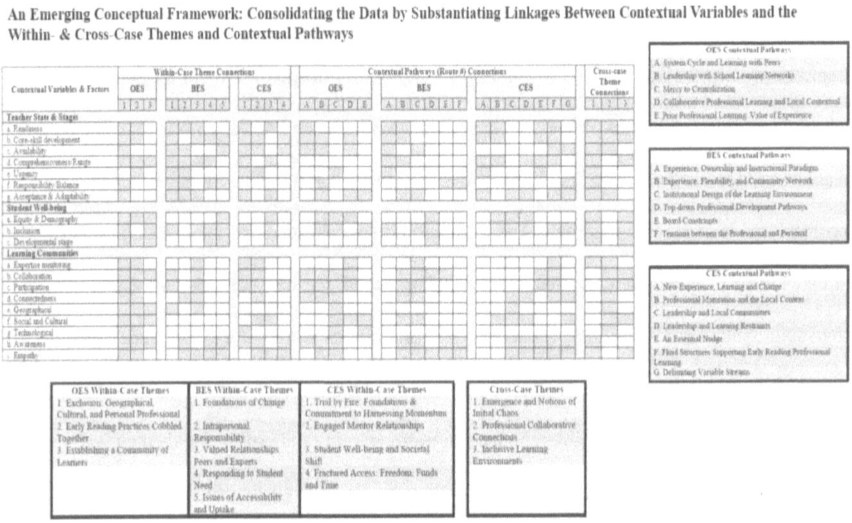

An Emerging Conceptual Framework: Consolidating the Data by Substantiating Linkages Between Contextual Variables and the Within- & Cross-Case Themes and Contextual Pathways

The learning, change, and achievement framework relates to how the self intertwines

between and within human relationships. These interrelations reflect the tensions between

awareness and considerations of more empathetically inclusive learning opportunities for

students across the education system. This framework conceptualizes the complexities

interacting upon this aspect of the teacher-learning system through a web of interacting and

individual notions of self to the contextual factors that interplay within individuals and the

systems with whom they are directly (and indirectly) linked. This interplay between self/selves

(student/s and teacher/s) and the professional community that teachers and students learn within

occurs between their relationships to/with the varying contextual factors brought into play in

their particular teaching and learning environments. Together, these contextual combinations influence how teachers perceive to have experienced learning, how context influences change, how teachers' and students' learning needs, and the system's accessibility and culture influences teacher change (see Figure 10).

Figure 10

Conceptual Framework: Learning, Change and Achievement

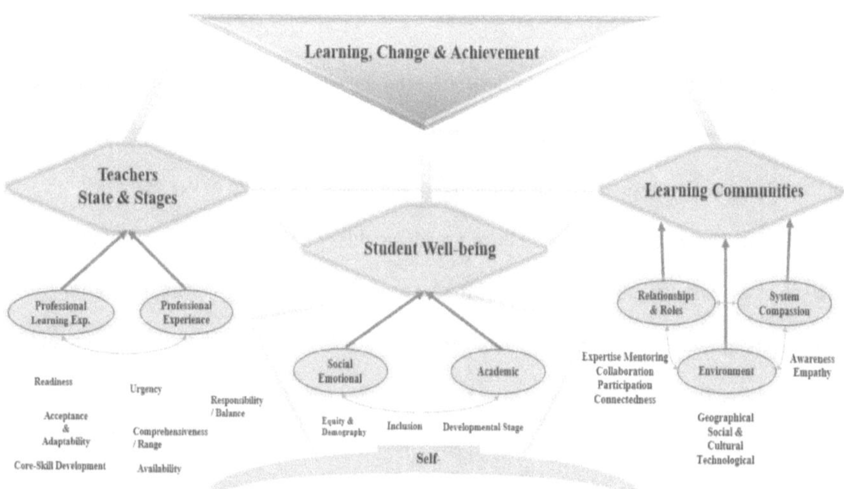

In this framework, particular notions of self are reflected in teachers and students as independent components working and learning within their educational social-systems. Together these notions of self are interacting within and amongst categories of contextual variables categorized by how they emanate between and relate more closely within one of three interrelated levels; the professional teacher, the student, and the broader system level, the learning community, that comprises the nature of professional teaching and learning and which includes principals, board administration and how the greater context influences professional

160

learning. These levels are at play and reflect a completely interconnected ecosystem represented by the variables and through the varying ways they can influence early reading learning and teaching and, subsequently, students' reading achievement. Considering integral aspects of teaching, learning, and the environment illuminates the importance of understanding systems as compilations of concepts of self and is a healthy way to consider notions related to professional learning designs.

The contextual variables are professional learning experience, professional experience, student social-emotional and academic well-being, roles and relationships within a learning community/system, a system's capacity for compassion, and the nature of the learning environment. These variables fall into three broad variables that reflect the notions of the self with selves as teacher/s, student/s and the broader professional learning system. Concerning learning, change and achievement, the three groups inextricably relate but opened up here in this two-dimensional platform they might appear distinct. They are not distinct, but still, each variable has its uniqueness. Figure 10 represents these unique factors for each variable. The conceptual framework provides a format that represents significant factors to consider in understanding the role of learning processes at play and influencing learning, change and student development.

Teachers State and Stages

Interacting with teachers' professional learning are teachers' states and professional learning stages, representing their professional learning experiences and overall professional teaching experiences. Teacher experience variables reflect the state of teachers' learning experiences and the stages of teachers' careers based on professional teaching and learning over time concerning their early reading practice and learning. Together, the factors interacting upon life as a professional teacher and directly connecting to the individual teacher are systematic labels and are fluid rather than fixed terms. More important is how meaning resonates through the names of the labels. Currently, seven factors represent ways to express the spectrums of interaction occurring within the teacher variables based on teacher participants' perceptions of early reading teaching and learning: (a) Readiness infers a sense of preparedness of the undertaking of teaching early reading. The spectrum of perceptions of being prepared (or not) to teach early reading effectively in different stages evolves through teachers' accumulated history and experience; (b) Core-skill development relates perceptions on the role of learning core teaching skills for teaching students foundational early reading skills that help students learn to read. Code-based instruction is a reading skill teachers believe needs to be taught to them and is a skill that develops over time from almost nothing early on; (c) Availability is critical to understand as specific experiences appear to be needed at different times in a teacher's career trajectory and reflects how much teachers value sustainability, which in essence is the meaning of availability; (d) Comprehensiveness/range is represented through the skills teachers need to have to teach all their students how to develop in their reading and so implies that to learn something the learning experiences have to be deep and comprise a range of learning experiences that different teachers need for developing the skills to teach reading; (e) Urgency is a separate factor included here because of how vital teachers consider it is to learn a range of teaching skills and knowledge early in their career and that this sense of urgency is embraced in perspectives relating a desire to learn a lot early on; (f) Responsibility/Balance comprises a dual nature of responsibility to the self-and-family and the self-and-professional, providing the right learning experiences for students reflective of the push and pull of balancing perceived personal and professional responsibilities; and (g) Acceptance and Adaptability infers degrees of flexibility to

162

take on objectives in the learning system such as a board supported early reading initiative. Though this list separates each factor, these seven factors should be considered holistically in how individual teachers vary based on their learning experiences and career stage.

Student Well-Being

The complexities of students' academic and social-emotional well-being are also central to the problem of learning, change, and achievement. Ideally, an education system's goal is to provide learning experiences conducive to developing core skills and strategies that help students become ethical, just, and critical consumers and contributors to society. In terms of student well-being, the contextual factors reflect values of equity and demography, inclusion and developmental awareness. Equity and Demography equate to shaping student learning experiences empathetic to students' history and current circumstances (e.g., addressing basic needs and rights to scaffold learning and positive development). These factors signify how the environment influences students individually and collectively, requiring particular teaching skills and empathy. Inclusion represents understanding how to shape specific learning experiences for students to include them as equals in the classroom. Developmental Stage implies the level/s of readiness students are at for learning specific skills in particular ways (e.g., appropriate instructional approaches or awareness of the stage groups and individual students are in).

Learning Communities

Interrelated within teacher and student variables are the complexities of learning communities. A Learning Community is considered here as a learning system and comprises three main variables: Relationships and Roles (expertise mentorship, collaboration, participation, and connectedness), Environment (geographical, social, cultural and technological) and System Compassion (openness, awareness, empathy, and learning-literate). Relationships and Roles refer to how a system organizes to form the learning relationships teachers respond to, informing their teaching. This variable relates to expertise, collaborative opportunities, options for participation and the connectedness in the learning experience. The variable, Environment, speaks to how the system or learning community reacts and adapts to needs in different environments to address and influence the learning and teaching process. Environmental factors are geographical, social and cultural, and technological. Geographical factors relate to proximal and situated space. Social spaces connect to physical and virtual places of learning and how social relationships influence teachers' learning experiences. Cultural factors closely connect to this but shift a focus to the influence of culture/s in the layers of the physical spaces within the learning community and how this influences teacher need. As a variable, System Compassion reflects the system's human nature and how it, as perceived as a range of interacting human intentions and tensions, influences the learning experience. Awareness connects to a system's openness and speaks to how a system provides and handles the varied needs teachers will have. Empathy connects to a system's awareness recognition, action, and capability to provide for the differences of all its learners (students and teachers) within its atmosphere.

This conceptual framework represents how the spectrums of such a range of factors influence teacher learning and student achievement. In the next chapter, along with briefly and sequentially answering the research questions, the conceptual framework shapes a discussion on teachers' state and stage, student well-being and the learning system. Following Chapter 5, the final chapter of this book concludes with implications for theory, empirical research and professional practice for positively influencing students' reading achievement.

Chapter 5 – Discussion

This qualitative, complexity-situated research study of teachers' perceptions of learning, practice and student reading development fulfills a need for understanding how teacher learning processes can enact practices to enhance the learning of all students, particularly those not well-served in the current system (Cochran-Smith et al., 2014). The goal of this study is to understand (a) how contextual factors occurring at varying levels are perceived to influence the delivery and uptake of early reading professional learning opportunities, and (b) how teachers perceive their early reading professional learning influences their early reading instruction and subsequently their students' reading achievement. This section begins by situating how I consider the concept of *struggling readers* in this research. I then briefly answer the two research questions in a linear like fashion before moving into a discussion of the findings structured across three levels: teacher, student and learning system.

Notion of Struggling

This research seeks to understand how professional learning opportunities influence teaching within the confines of the 21st-century educational learning and teaching system. Chapter 2 indicates that the cognitive, psycholinguistic literacy paradigm is more influential in common conceptions of Western education's English Language Arts reading instruction. The focus of this research seeks to understand participants' perspectives regarding literacy instruction and early reading professional learning. For the teacher participants, this mostly relates to code-based instruction and setting up the learning-to-read environment. Underpinning this research is the notion of who a struggling reader is. Teachers at the rural and urban schools perceive that challenges in learning to read often centre on students' sociocultural context as inhibiting their opportunities to learn and how they and their school can help create atmospheres sensitive to

165

their unique context. Teachers at all schools also consider that struggling readers have a disability or difficulty ingrained within that makes it hard for them to *crack the reading code*. Collectively, notions of struggling and at-risk imply that a student is not meeting the school system's measures of academic success. This research seeks to understand how teachers' learning context influences their learning, their teaching, and the success of all of their students. What is clear is that when someone does not fit into this system, they will be considered as unready or unequipped compared to those for whom the system is a better fit and easier to be successful in.

It is evident that the teacher participants care for their students and want nothing more than for them to learn how to read so that they have the foundations that set them up to have the freedom to contribute within a society that places so much value upon people's ability to read printed text. Further, the case schools in this research appear to be taking steps to shift and prepare their students to become more literate. I have the perception that this is scratching at the surface of inclusive early literacy practices but believe that the findings in this research illustrate how the professional learning system can adapt to be more inclusive so that notions of *failing*, *struggling*, or *at-risk* relate to a system's deficit and not a human being's deficit. In this way, varying literacies can be a part of how we celebrate success and consider learning to read beyond the traditional modes of reading achievement societal institutions use as passcodes for entry into positions of positive societal influence.

The Short Answer

This research seeks to understand the professional learning system's role upon early reading instruction and student reading success. It situates itself in literature demonstrating how a large proportion of beginning readers continue to confront immense struggles learning how to read even with a substantial body of research expressing (a) how students should learn to read,

166

(b) what elements of early reading instruction effectively improve students' reading and (c) ways professional learning can improve teachers' early reading knowledge and practice. Considering that many students struggle to learn how to read in the critical early years (e.g., Allington, 2011; Schuck et al., 2018; Seidenberg, 2017; Torgesen, 2002; Wolf, 2018), my research, in utilizing a complexity theory qualitative case study approach, explores the relationships between teachers' professional learning systems, their teaching practices and their perceptions on how their teaching influences student reading development. By considering how the system influences teachers' and students' work and learning, this research addresses a gap between what the research suggests and what is happening for many students when they learn to read. It is an important topic because of the social inequities poor readers confront, and this research is needed to help inform and re-calibrate current professional learning systems to inform how education can better meet all of our students' early reading needs. To achieve these goals, the two research questions leading this research study are:

1. How do contextual variables at the school, board, and provincial level influence the planning, delivery and uptake of early reading professional learning opportunities?

2. How do teachers perceive the relationships between (a) their professional learning experiences, (b) their classroom early reading practices, and (c) student reading outcomes?

These questions have one overarching purpose; to develop an extensive sense of how context, teacher learning, teacher practice and student reading interrelate. Together these questions provided the entry point for delving into teachers' perspectives of their learning system/s and result in a framework for conceptualizing an equitable and inclusive system for early reading education and professional learning, or teacher learning and student achievement.

The first research question asks, "How do contextual variables at the school, board, and provincial level influence the planning, delivery and uptake of early reading professional learning opportunities?" This question's answer will follow the chain of command structuring educational institutions: province, school board, and school (principal, teacher, students). Varying strengths of interacting factors emanate from each system to influence how the professional learning system interacts with teachers' early reading professional learning.

Province. Provincially, the Québec Ministry of Education is responsible for establishing the school system's curriculum and structure across the province. This research did not delve deep into the curriculum structure; however, Chapter 2 demonstrates there is not a clear description for teachers to follow for students' early reading instruction. Additionally the board-level participants in this research indicate the challenges teachers are having in terms of implementing it. In terms of structuring the system, the most considerable influence that the province has on teachers' uptake of professional learning appears to be the role of the cycle grades in engaging teachers with their cycle peers.

Having colleagues in the same cycle ensures degrees of professional conversations as students develop subject competencies over two years. Teacher participants from OES and BES value this structure for varying reasons. OES's isolating context limits the opportunities for OT1 and OT2 to go beyond their school community, and the cycle system affords them a form of colleague companionship. Both highly value this collaboration in their sparsely populated school. OT1 and OT2 crave collaborative learning experiences, and for them, the cycle system is one outlet that affords them opportunities to develop, accumulate, and share reading resources and instructional strategies. BT1 and BT2 also value the professional early reading conversations they have with their cycle partners. They appreciate how having cycle teams helps them incorporate board-supported reading initiatives into their professional conversations at school. Interestingly, the cycle system was not a part of the conversation with teachers from CES. One additional, minor, though still influential aspect to the province were concerns from participants on the timing of provincial professional learning days. Participants mention how these days often fall near holidays and reporting periods and that this does not make it desirable for them to attend.

School Board. The second level in the hierarchical chain of command within this education system is the school board. In this research, teacher participants highly value it when the school board designs professional learning opportunities that connect teachers from similar contexts and provides a reading expert to visit classrooms. Teacher participants prefer board supported early reading programs that sustain over time, provide opportunities to collaborate with peers, are led by mentors and include opportunities for teachers from different schools to connect and establish lasting professional relationships. Mentors and experts are highly valued. Teacher participants attribute their early reading knowledge, instructional strategies, career longevity, current role as a model teacher and their successful reading programs to their relationships with particular mentors.

However, when board-supported early reading programs do not connect with the school's local contextual needs (e.g., language at OES, social-emotional at CES), teachers do not overly value what the board offers. Additionally, school boards face challenges in providing isolated schools with practical, relevant and engaging technologically mediated professional learning. This research indicates how unfavourable and unstable OES participants perceive web conferencing is. For OES, the context of their isolation is a factor that their school board needs to address to improve professional learning.

Barriers to professional learning that teacher participants perceive in this research reflect cost, timing, topic and an overall sense of reduction in early reading professional learning. Specific to the two teachers at OES is their isolation, and the personal time and money they have to spend to attend centrally offered professional learning. Teachers who are mothers face exclusion from the learning communities they are accustomed to when they take a leave to have children and upon their return to teaching and have numerous obligations to their families after school hours.

Finally, one resounding belief expressed by all teachers is that there is no professional learning provided from their school board to help them learn how to help students who struggle to read. Specific difficulties generally revolve around general conceptions of students finding reading hard or students having learning difficulties. However, in these conversations, participants perceive that they do not have the skills to help students who struggle and that the board is not providing them with any opportunities to learn how to support their struggling readers.

One barrier influencing the delivery and uptake of early reading professional learning reflects the school board's role and includes the principal to varying degrees. Above, the role of physical exclusion in negatively influencing OES teachers' perceptions of professional learning expresses the importance of understanding how to connect with teachers and including them in an effective professional learning process. Participants are wary of how to access some learning opportunities and believe that the school board often controls who will attend. Further, participants are unsure how to obtain funds or access to professional learning opportunities and that the board consults principals' opinions and recommendations of specific individuals. The perceptions of the participants who voice this concern indicate that there needs to be a more transparent process for teachers in their professional learning aims, opportunities and directions. Otherwise, as teachers perceive, a sense of distrust and disengagement can materialize.

An interesting finding relates to teachers' perceptions of how influential the teacher induction program is that all new teachers to the board take part in. Notably, no teachers consider that this program is influential upon their reading instruction. Participants would like to see these mentor-mentee relationships more relatable to their early reading needs as beginning teachers. There are potentially significant benefits of a teacher induction program, yet this does not resonate with teacher participants in this research. Understanding the emotional experience of becoming a new teacher and combining it with research on teacher knowledge levels can shape possibilities for creating practical ways to structure new possibilities that connect with these teachers' emerging early reading needs.

School Level.

Principal. Principals influence the uptake of early reading professional learning in several ways. First, the school's sociocultural context appears to influence the choices principals make towards professional learning. CES' principal embeds school-led learning with workshops and discussions on attachment theory. CP believes that unless students in the school learn to view the school as a safe place, they will struggle in all facets of school life. At OES, OP organizes school-based professional learning around developing their predominantly French-speaking population's English language capacity. BES does not have a comparable population of students to the other two case schools. BP works with the staff on visible learning and developing students' metacognitive skills. Secondly, relating to the point above are the professional perspectives underpinning principals' beliefs about how students learn in their schools. For example, this appears to influence choices they make concerning who teaches specific grades and what type of teachers principals want to hire. CP considers grade 1 the most critical year of elementary school for students and so the teacher in that role is someone he believes will continue to seek ways to help their students learn to read and regulate: and this appears accurate when considering the path CT2 is on concerning her early reading development and role as a model early reading teacher within the board. OP is an experienced rural school teacher and principal, and part of her position goes to teaching. Along with a strong literacy instruction background, OP has her finger on the pulse of her students' day-to-day needs and is adamant that she will not be someone who leaves the classroom to be a full-time administrator. BP believes struggling readers remediate over time; that difficulties with reading relate to the speed of development and that by early high school, most students will grow out of this problem. The varying personalities of each principal influences the options for early reading professional learning at the school level.

Teachers. Teachers' level of experience profoundly influences their perceptions of their participation in early reading professional learning. Teacher knowledge develops over time, and setting the foundation through a range of experiences at the beginning of teachers' professional careers is essential in their early years. The four most experienced teachers in this research consider their early years overwhelming but upon reflection, acknowledge the importance of participating in professional learning that prepared them to provide core cognitively based principles of early reading instruction. Context strongly influences perceptions of teacher-learning needs, and though teachers are receptive to some elements of the school board's early reading professional learning, the board needs to align with teachers' specific sociocultural situation. Four participants explicitly declare that they take pieces of the most recent reading programs supported by the board but will not fully commit to these programs or the associated professional learning because it is not contextually relevant to their students' needs. Preparing beginning teachers for the abrupt entry into their profession, along with information about the general developmental trajectory teachers traverse across their career, will help teachers realistically discover and broadly map out their learning paths. Further, understanding how layering professional learning overtime influences instruction, teachers will benefit from a system that differentiates to teachers' varying strengths, provides collaboration, offers mentoring, seeks teacher mentors, and creates social learning opportunities for teachers within and across schools. After briefly answering research question two, the discussion below expands upon these types of learning systems.

Research Question Two

The second research question asks, "How do teachers perceive the relationships between (a) their professional learning experiences, (b) their classroom early reading practices, and (c) student reading outcomes?"

Professional learning experiences that comprise multiple and varied learning experiences early in one's teaching career involve collaborative learning opportunities with peers, reading experts and mentors and relate to the context of teachers' work. These elements translate to perceived positive instruction and improved reading outcomes for students. Describing the notion of achievement and student reading outcomes at the onset of this book, I make the point that this is not reading achievement per standard normative based assessments but rather, concerns how teachers perceive that their instruction helps their students become confident and fluent readers.

Teachers who receive a range of early reading professional learning early in their careers use it to anchor their reading instruction and future professional learning. To beginning teachers, this does not appear to be a self-evident revelation when they are in the early and abrupt process of becoming a teacher, but it does appear to inform their reading instruction across the years when they have the time to reflect on this process of learning. An openness to a range of early reading professional learning also relates these teachers' receptivity towards shifting instructional ideologies. For example, in this research, three teachers who began their career teaching from a whole-language perspective speak about incorporating more code-based instruction after experiencing it in professional learning and improving their practice and their students' reading progress. These teacher participants are aware of how culture influences their students' readiness to learn to read. Specifically, teachers from CES and OES value professional learning that connects to their students' sociocultural characteristics. Professional learning that reflects their students' identities reverberates with these teachers, informs their practice, and engages their students. Mentorship is also integral to the development of teaching early reading instruction perceived to influence their students' reading success. A perfect mentor supports teachers' confidence, provides effective instructional strategies for early reading instruction, and demonstrates how to deliver effective instructional strategies and reading programs and is available on short notice via emails, phone, text, and personal visits to the classroom.

The above discussion is fairly surface level and communicates a succinct culmination of the extensive data analysis in a linear-like response to the two research questions. It is useful as a starting point for setting into motion some broad overarching elements and avoiding certain variables that act as barriers to teachers' early reading professional learning and student reading achievement. The next section provides a lengthier and contextually more significant discussion framed within the main variables interacting in the teacher learning, change and achievement process: the student in the centre supported by the teacher on one side and the learning community on the other.

Learning, Change and Achievement

This section answers the research questions using the conceptual framework presented at the end of Chapter 4. By focusing on the teachers' lived experience within early reading and professional learning, I have explored an under-considered aspect to early reading research by utilizing complexity theory to untangle how contextual factors influence the variables at the front and centre of reading instruction: teachers, students and the learning system. Framed within complexity theory, this section of the discussion connects the findings to relevant research on teaching, learning and early reading instruction, situating a way for conceptualizing early reading instruction as an emergent, self-organizing and nested process dependent on a learning system adaptive to the varying teaching and learning needs it comprises. Conceptualizing the interacting factors upon and within the variables of teacher state and stage of learning, student well-being and the learning community, an inclusive professional learning system for teachers and students emerges. The findings weave a complex web of how the contextual variables influence teachers' perceptions about meeting their students' early reading needs. Within this human social learning collective, the contextual factors at play and discussed here are highly influential upon each of the variables' ecosystems. Conceptualizing the learning, change and achievement process as a more elevated interconnected ecosystem than its parts provides an inclusive pathway towards a decentralized learning collective explicitly valuing diversity, compassion, and individual differences. Following this complexity situated discussion, Chapter 6 concludes this study suggesting theoretical, empirical and practical implications and the limitations of this research.

Teachers: State and Stages of Professional Learning and Experience

Results in this research indicate the receptivity teachers have towards undertaking a range of early reading professional learning and development opportunities in their beginning years. Over time, teachers' nested nature within a particular stage of their career represents different early reading professional needs and how developing comfort, confidence, and a philosophy of reading instruction emerges slowly and through a range of learning experiences. In line with sociocultural theory on teacher professional learning, effective collaborative learning experiences often, though not always, provide a type of community that nurtures teacher identity, knowledge and practice (Kelly, 2006). The data also presents how visible, and invisible forms of complexity reduce and redirect teacher learning opportunities and their practices and students' early reading experiences. Complexity reduction "refers to ways in which the complexity of social systems may be reduced" (Hetherington, 2013, p. 74). In terms of the teacher variables in this research (state and stage), complexity reduction is visible in how ideal conceptions of learning are perceived to be present (or not) and how these perceptions reflect the emotional sensitivity in terms of fitting in, self-confidence and beliefs around having needs provided. Complexity reduction occurs across all stages and nested levels of teachers (ages, schools, experiences, personal lives) by limiting teacher learning and practices in similar and varying ways throughout teachers' careers.

Teachers with varying levels of experience have different learning needs, and early career teachers need access to a range of opportunities for learning how to teach reading and to develop critical foundations to their professional practice (e.g., Fletcher et al., 2013; Hargreaves, 2005; Mahmoudi-Gahrouei et al., 2016). Emphasis on understanding the emergence of self-organizing systems needs to be better considered concerning teachers' experiences and needs and the new complex systems they enter as beginning teachers and while they progress along their professional trajectories (Cilliers, 2000; Morrison, 2008). The range of the career and life stages of the participants in this study provide data demonstrating a spectrum of teachers' professional needs. One constant finding this research highlights is the value that all participants place on their learning community. For example, in the context of the Québec system, the cycle partner is invaluable to most teachers. Though she is critical of the learning opportunities in isolated OES, OT1 finds her cycle colleague an invaluable professional companion. BT1 and BT2, along with other teachers at BES, communicate via a shared, rich professional language throughout the hallways, via impromptu meetings in each other's classes after school and during lunch in the staffroom. The power and desire for collaboration with school colleagues emerged in conversations with BES and OES participants in various strengths and with varying effects and regulating the development of a shared school language. Collaborative forms of learning are an essential part of gaining access to a community and developing a professional identity.

Teachers' professional identities jolt into action in their early years (Hargreaves, 2005; Hetherington, 2013). Individually, within their new system/s of learning, beginning teachers seek to self-organize in ways for teaching practices to emerge, which helps them understand basic constructs for teaching early reading and meeting their students' needs. It is an initial developmental process that the teachers in this research undergo, and that appears to occur intensely in their first few years while they adjust to a dizzying and overwhelming professional environment. The two beginning teachers from OES share a current snapshot of this experience as they discuss their uneven processes of cobbling together their early reading practices. For example, OT2, in her second full year of teaching, embodies the hectic life of the beginning teacher. She reacts to immediate needs instead of longer-term unit and lesson planning and admits to planning twenty minutes before the school day begins because she just knows where she is going. Overall, the four most experienced teachers in this research (11-21 years) relate how meaningful their variety of early reading professional learning experiences in their early years are on their teaching. For them, these are critical times to experiment and learn about different elements of reading instruction. Upon reflection, BT2 views her professional learning experiences fitting together like a puzzle even though this was not the case in her beginning years during the scramble to develop an identity as a professional teacher. Interestingly, OT1, in her third year of teaching, also alludes to the notion of layering past professional learning to enhance her reading instruction.

These teachers perceive that their early professional learning experiences are foundational to their current core early reading practices (e.g., being introduced to the elements of guided reading instruction, setting up their classroom environment for reading, and adapting their instructional beliefs). Their experiences have been catalysts to their existing knowledge and reading instruction by helping them become flexible in their approaches and providing practical tools such as technology (e.g., Smartboard) and program structures (e.g., guided reading). Likewise, Hargreaves (2005) finds that experienced teachers consider that their early years are crucial in their development, where an eclectic range of professional learning opportunities is welcome. His research demonstrates teachers awash in strong emotions (fitting in, striving to be good teachers) while they often, and abruptly, transition into their new roles and responsibilities. In this kind of frenzied atmosphere, it can be difficult for new teachers to make professional learning choices instrumental to their early reading instruction. CT1's and CT2's acknowledgement of making some ineffective choices in their first few years of teaching exemplifies this. However, they admit that some learning opportunities, such as the Bureau of

178

Educational Research, influence their current reading instruction, and larger populated one-off professional development productions do not leverage their professional capital.

Importantly the findings indicate that the rural teacher participants, though appreciating elements of their professional learning, perceive that much of the early reading professional learning available to them are void of instruction on more core individual teaching strategies. This finding does not necessarily represent negative perceptions of the professional learning the two teachers from OES receive but indicates that both teachers believe more opportunities to learn how to teach the early reading skills that their curriculum requires will help them develop upon the basic skills they perceive they possess. As it stands, OT1 and OT2 represent teachers in their early stages of developing a sense of self-efficacy to meeting their students' early reading needs. Research supports how this appears to be the most critical stage to learn focused and wide-ranging skills. Hargreaves' (2005) findings on teacher change throughout a career reflect similar notions to the two beginning teachers from the OES, the rural case school. Teachers early in their careers value and need a range of learning experiences that will ground some of the core components of teaching reading. Concerning the self-organization of beginning teachers, practices emerge and develop across time. Therefore, these first few years of teaching are the ideal stage of their careers for teachers to learn a range of methods and skills to provide reading instruction and put things into play (practice). However, it also appears that beginning teachers have trouble understanding how to synthesize experiences, and only years later, when reflecting on their early years, do they see how early reading professional learning imprints upon their knowledge. Along with developing a core knowledge base, the beginning years are also a time for new teachers to find their places as professional teachers.

All teacher participants in this study perceive the influence and value of early reading professional learning upon their practice and students' reading development and success. Teachers' early professional learning experiences equally influence their sense of initial grounding towards a more in-depth perspective of their professional sense of self. The early years stand out as a junction in a teacher's career when they need professional learning sensitive to their readiness and potential. It is also a time to help anchor their understanding of early reading instruction; a time and place when foundational beliefs and practices are emerging and malleable and when the introduction of specific practical skills are necessary (Day & Gu, 2006; Mahmoudi-Gahrouei et al., 2016; Pillen et al., 2013; Schuck et al., 2018; Van der Klink et al., 2017). Participants often noted the influence of and desire for learning with early reading experts. One existing learning opportunity for all teachers beginning in the board is the teacher induction program. Research reflects how challenges new teachers face are often not worked out during their induction and mentorship experiences (Schuck et al., 2018). Even when beginning teachers have mentors and smaller classes in their first years, the problem areas not resolved persist years later, emphatically acknowledging the importance of considering new teachers' needs in ways that teaching development priorities match theirs and their students' learning needs. Interestingly, teacher participants' do not perceive their induction experience influenced their early reading instruction. School boards with a better understanding of their beginning teachers' early reading knowledge will have some critical information to design learning opportunities early in these teachers' careers to help them develop a solid foundation for teaching reading to all of their students.

Pillen et al. (2013) identify the importance of fitting in as a beginning teacher and the value of professional learning in these early years to help decrease professional self-doubts by increasing perceptions of professional and personal self-efficacy. Kelly (2006), viewing teacher education through a sociocultural lens, stresses how critical it is to understand social learning processes and experiences beginning teachers need so they believe that they are part of the professional community they are entering. New teachers are on the periphery of their profession and require collaborative experiences to help learn content and develop their identities as teachers. The diverse environments of participants in this study (rural, suburban, urban) represent the role sociocultural factors have on developing a teacher's professional identity in different teaching environments. In this research and reflected in Hargreaves (2005) and Pillen et al. (2013), it is clear that a strong focus on developing core skills can develop practical and core teaching strategies and improve teachers' self-efficacy and self-perceptions in their teaching community; and that this contributes to a sense of belonging.

Further, and concerning complexity theory, the nature of connectivity and openness in systems is enhanced by professional learning that empowers many new teachers' core skills to develop and positively affect their professional self-efficacy and sense of professional community (Cilliers, 2000). Here, the notion of adaptiveness of complex systems is central for empathizing with the developmental process of new teachers' identities within nested systems (Davis & Sumara, 2005; Morrison, 2008). For the most part, the findings on the nature of being a beginning teacher relate the sense of urgency and excitability that new teachers experience, longing to carve out a breadth and depth of professional knowledge. However, mediating these perspectives, particularly for the rural school teachers at OES, is exclusion and how being nested in a rural school can shape attitudes towards their professional learning system. The three participants from OES share the challenges that their geographical and professional social isolation has on their learning, perspectives towards learning communities and overall sense of connection with other less isolated teachers and schools. For example, OT2 relates their school's physical and social location akin to being on the edge of the earth, and OT1 and OT2 worry that their isolation invariably impedes them from leaving to obtain professional learning beyond their school. Their particular situation represents educators on the outside looking in, while simultaneously, their peers in less excluding environments can seamlessly integrate into diverse and active collaborative learning communities. This sense of exclusion to collaboration extends into virtual spaces as well. Internet connectivity is not reliable, and currently, OT1 and OT2 perceive web conferencing severely lacks the personal connection they believe face-to-face collaboration provides. Teachers at BES similarly discuss the value of collaboration. However, unlike their rural counterparts, they have access to more teachers in their school and are more centrally located. The value teachers at BES place on learning with their peers accentuates the importance for teachers to have access to environments that nurture a shared professional language and use the diverse knowledge and experiences larger collectives of teachers offer each other and, therefore, a place for emergence of knowledge and sound reading practices.

BT1, BT2, CT1 and CT2 consider specific professional relationships highly influential in their transformations in becoming fully-fledged members of the teaching community. For example, CT1, who nearly quit teaching in her second year, experiences a perspective-shifting metamorphosis following a rigorous one week in-situ reflective, collaborative experience with a highly experienced educator. For CT2, her relationship with SB3 beginning in her first year teaching grade one is the impetus to her participation in the school board's early reading professional learning program, and in which she is now a model teacher. BT1 considers two very early mentors prominent influences upon her early reading instruction. Along with developing core early reading skills, this experience was responsible for diminishing negative self-perceptions and increasing her professional self-efficacy. It is apparent that when teachers can work alongside knowledgeable and dedicated professionals, there is a synergy that ignites positive change, practice and beliefs. OT1 and OT2 unfortunately, do not mention similarly profound, influential experiences, but because they are in the early stages of their careers, they will hopefully have similar experiences.

Understanding the nature of nestedness is an integral component to conceptualize teacher identity, early reading belief system, and emergence and self-organization as a professional. Davis and Simmt's (2006) discussion and depiction of nested complex phenomena present a timescale for considering the relationships and transformations teachers move through developing varying orders of knowledge. Their findings connect to a perspective on how novice teachers transition over time to become experts (Kelly, 2006). Citing social theorists Shon, Lave and Wenger, Kelly (2006) iterates how becoming an expert teacher requires opportunities for engagement with other professionals to help transition novices from a peripheral position of learning towards full participation: "In their movement from novice to expert, people adopt different stances towards the task in which they are engaged … Facets of such constructions include how teachers interpret their role, the meanings and understandings which they bring to their role, their beliefs and intentions" (p. 513). Understanding beginning teachers as on a periphery underscores an urgency for others with more experience and knowledge in the community of professional teachers to help bring them into circles to share, learn and develop their teacher identities.

The personal nature of a teacher's life nests within the personal and professional influences encompassing them, interacting to influence their access to professional learning and development (Avalos, 2011; Hargreaves, 2005). Whereas at the beginning of their careers, all

183

teachers are entering a community for the first time, life situations such as family obligations and responsibility are also interacting and disrupting how teachers access their professional communities. These varying needs comprise an element interplaying with the professional learning experience. BT1, BT2, CT1, and OT1 report how economic interests conflict with their professional learning and diminishes their likelihood of seeking professional learning. OT1 perceives it is often too expensive to pay upfront for professional learning when the returns on knowledge gained are minimal. For the three teachers with children (BT1, BT2, and CT1), priorities for spending belong to their families. The commitment to raising a family places a firm boundary on most primary school teachers' professional growth for sustained periods in their careers. Considering that 84% of the elementary teaching population is female, too many teachers at varying points in their careers relinquish the learning communities they depend on to change practice and influence perspective (Statistics Canada, 2017). It is also very likely that teachers who have to step out of their professional community to raise their infant children never quite regain entry for various reasons. For example, BT2, whose child is school-age, does not consider professional learning that will take her away from her family. However she does attribute early reading professional learning occurring in her school to be instrumental for learning effective strategies to manage her instruction. BT1 and her husband (who works nights) are raising several children. Only very recently has she more time to attend some professional learning because her mother moved back to Blending. However, this is likely not the same scenario for many teachers who are mothers and struggle to balance their personal and professional responsibilities.

Davis and Simmt (2006) depict subjective understanding nested in classroom connectivity, nested in curriculum structures, nested in mathematical objects as categories of nested knowledge. Subjective understanding and knowledge can happen over a matter of seconds, while conceptual knowledge of mathematical objects develops across decades of intentional participation. Their breakdown of learning timelines relates to identity and connects to the findings in this research regarding the epistemological assumptions teachers hold as they enter their profession. Important, though lightly touched on in this research, is the role of shifting foundations and emerging instructional beliefs teachers go through. For example, beginning their teaching careers with a preference for providing a whole reading environment for reading instruction, BT1, BT2 and CT1 undergo a shift in their approach to reading instruction following professional learning on code-based instruction early in their careers. Interestingly, a range of support for whole-reading, code-based and balanced reading exists (see literature review in Chapter 2). However, literature over the last two decades primarily promotes how necessary it is for teachers to implement scientific cognitively-based early reading instruction to teach students to crack the reading code (Torgesen, 2002; Wolf, 2016). Thus, it is not overly surprising that participants mention their shift to code-based and nearly no emphasis concerning other forms of reading and literacy. At the time of this research, OT1 and OT2 are searching for ways to help their students crack the code. Understanding their entry into their profession as a process of forming an identity from the periphery into full membership indicates that providing a range of more code-based learning experiences will help speed up this integration, but also recognizes that including more contextualized notions of literacy will make learning to read more inclusive (Kelly, 2006). The emergence of a teaching belief system, or teachers' philosophical and ideological approach for conceptualizing early reading instruction, is a factor that requires nurturing over time; and included in this is expanding ways for including more students to incorporate more naturally culturally relevant literacies.

In terms of literacy paradigms, teacher self-belief is also a critical factor in the teaching-learning process. A brief overview of literacy paradigms in Chapter 2 presents two main and varying views on literacy and their relationship to reading instruction. A cognitive, psycholinguistic paradigm on reading privileges reading instruction that provides code-based instruction where students learn letter-sound relationships, moving stepwise to becoming fluent readers of longer text and developing reading comprehension of printed material (Beach & O'Brien, 2017; O'Brien & Rogers, 2015). This view prevails in most elementary schools and teacher education programs (Beach & O'Brien, 2017). Very broadly speaking, sociocultural literacy paradigms focus on situated and contextual natures of literacy through considering the value-based implications of literacy (meaning from print, meaning from and across multimodalities and meaning from reading the word and the world) and how the mainstream society and institutions do not leverage the literacies oppressed and less advantaged individuals and cultures value (Perry, 2012). Connecting code-based strategies to student reading improvements is a traditional way to consider reading instruction but largely omits sociocultural considerations for why too many students struggle to read (Dillon & O'Brien, 2019; Perry, 2012).

Most participants commented that their bachelor of education program did not effectively help them learn to provide code-based instruction, instead preparing them to provide engaging literacy-rich whole language atmospheres. This foundation of this preservice learning environment raises a critical point. English language arts preservice education does not appear to provoke, at minimum, memories of a critical sociocultural aspect to the nature of literacies in these programs. In turn, this influences the ways teachers consider how to deliver reading instruction and prioritize their professional learning. Together this symbolizes how current structures privilege a cognitive (code-based) approach. For example, no teachers overtly mention

186

this beyond an attempt to connect schooling to meet second language and attachment needs (OES and CES). Ultimately this perpetuates a perspective that the many students who struggle to read have a deficit within them that needs repairing rather than a system in need of reform (see Perry, 2012). However, a ray of optimism exists as the four most experienced teachers experienced shifts in integrating code-based and whole language approaches, and CT1 and CT2 place a priority on adapting instruction to their students' contextual uniqueness. Together, this suggests that the participants in this research can likely expand their literacy conception as more inclusive to students with rich and strong literacies (Beach & O'Brien, 2017; Perry, 2012).

This section above presents a cluster of interrelated contextual factors influencing professional learning and teaching experience. Interacting and interrelated factors within teachers' early reading learning and practice reflect degrees of teachers' readiness, range and quantity of experience, learning hoped for, an intrinsic ability to shift philosophical beliefs and expand knowledge over time, and the influence of collaborative experiences in transforming teaching. This research study vocalizes teachers' desires to have professional learning inclusive towards a school's cultural context that provides diverse topics that relate to literacy and learning to read, and includes regular opportunities to collaborate with experienced peers. Says Kelly (2006), these kinds of

> explorations would be specific to particular groups or individuals and the context of their practice … [I]t is the constant and iterative engagement in constructing and reconstructing professional knowledge using various perspectives including teacher research with the aim of conceptualizing and addressing problems. Here teachers have an active and productive relationship with their professional knowledge base. They construct their own knowledge base for teaching, in their own particular circumstances… (p. 509)

Teasing out the contextual factors interacting upon teachers in this research demonstrates that teachers want to collaborate, value mentoring, seek knowledge, need to have their personal lives considered, and importantly, are flexible in shifting beliefs when some aspects for learning are available to help them learn, practice and change. All teacher participants value learning with others, but for beginning teachers, this also offers an opportunity to be included in their new community. At the centre of the teacher-learning experience, and discussed next, is student well-being.

Student Well-being: The Academic and Social-Emotional Combination

In part, the previous section discusses the importance of addressing teachers' core reading needs in their early years to help provide a foundation from which to develop and scaffold teachers into their professional teaching communities; both critical elements in the construction of a teacher's identity and self-efficacy (Hargreaves, 2005; Kelly, 2006; Mahmoudi-Gahrouei et al., 2016). Beyond the notion of the developing professional self, it is essential to understand how teaching knowledge and skill influence students' reading experiences and successes (Allington, 2011; Castles et al., 2018; Fletcher, 2013; O'Brien & Rogers, 2015). The act of teaching and learning is a social process between and amongst students and teachers and requires a combination of empathy and knowledge from teachers towards their students (Kelly, 2006; O'Brien & Rogers, 2015; Perry, 2012). Findings in this research indicate how crucial teachers perceive it is for them to meet their students' social needs and how social skills have to be in place to be academically ready for learning. Complexity theory frameworks denote the importance placed on understanding nestedness, self-organization and emergence of effects influenced in layers and tensions of interacting expectations, responsibility, diversity and beliefs (Morrison, 2008; Davis & Sumara, 2005; Osberg et al., 2008). For example, consider how a struggling reader is nested within a class who has a teacher unable to provide instruction to leverage students' strengths and abilities. This could be a teacher unable to offer code-based instruction due to faulty learning experiences, or a teacher who cannot provide an engaging and safe atmosphere for students to take risks. It could be a teacher who is not culturally cognizant and socially aware of how their teaching does not connect to their students' diverse backgrounds. The ways this might compound into a crisis with the wrong combination of teacher/s and student/s is troubling and very possible.

This multiple case study of a rural, urban and suburban school's contextual uniqueness provides a panoramic of how local natures and students' needs are often top of mind when teachers relate to their professional learning perspectives and needs. For example, teachers at OES work incredibly hard to connect to their students' French language backgrounds. At Blending Elementary, BT2 attends attachment workshops to help her create a restful and safe learning environment. All CES participants are mindful of the many students' in the school who have challenging home experiences, and they take steps through professional learning to incorporate this awareness into their school and classroom practices. Catalano et al. 's (2006) research on school connectedness (school bonding [attached relationships to school] and commitment [investment to doing well in school]) reveals the importance of establishing close student-with-teacher and student-with-school relationships because low connected students are more likely to abuse drugs, be delinquent, commit a crime, and drop out of school. Other research indicates the relationship between students' school connectedness and academic success and environmental factors related to the home, such as socioeconomic status, attitudes to learning, and cultural priorities (Forget-Dubois et al., 2009; Noble et al., 2006). In turn, and over the long term, these attitudes are a factor in student learning across time and are already developing when children enter school. Detached attitudes from students towards school are a disposition more urban and rural school teachers will face, and the learning system organizing professional learning needs to consider this (Cochran-Smith et al., 2016; Grudnoff et al., 2016).

This line of thinking relates to my research in how relationships (teachers and students) influence the professional self but also in the ways that contextual factors shape the needs of students and create the emergent needs unique to particular locations. In this research, struggling readers are perceived to be students who have a learning difficulty (OT1, BT1, CT1, CT2), whose first language is not English (OT1, OT2, CT1, CT2), who are low socio-economically (CT1, CT2), or who are First Nations (CT1, CT2). It appears that these students, unfortunately, struggle more than others, and these struggling readers represent a substantial portion nested in the grades in their schools and across our society (Allington, 2011; Frontier College, 2017, 2018; Government of Canada, 2015). Because the principal of Central Elementary perceives how challenging academic expectations are for many students, he strives to provide teachers at CES professional learning to create safe and motivating classrooms and prioritize learning, which puts the notion of whole student well-being in front of the formal academic benchmarks teachers are responsible for reporting. CT1 and CT2 respect CP's outlook on connecting to the student body, taking part in the related school-based professional learning, and seeking learning opportunities beyond board-supported early reading professional learning; though not all CES teachers follow CP's community-based approach to schooling. Relatedly, all participants at OES admit to shifting their more traditional elements of teaching English Language Arts to connect to their students' sociocultural differences in the predominantly French-speaking lower SES community of Extére. In general terms, emergence materializes through patterns of self-organization (for example, what teachers learn and board or provincial guidelines for what is acceptable numbers of properly achieving students) and reflects how teachers might or might not transform their teaching in systems that expect specific indicators of achievement. Whereas teacher participants from CES and OES self-organize their learning and teaching by incorporating social-emotional and cultural aspects into their teaching, this same belief in changing practice is not as urgently explicit in the mostly middle-class Blending.

The findings on perceptions of teachers towards struggling readers indicates that students continue to face challenges learning to read because professional learning opportunities generally do not exist to prepare teachers to help these students learn how to read academically. Though this troubling state of affairs opens up questions about how particular literacies are privileged, all participants in my study acknowledge the low number of professional learning experiences they have had to prepare them to teach students struggling to read printed text. This recognition represents how complexity for learning to teach reduces teacher learning experiences by tilting towards reading instruction catering to the needs of an exclusive group of students who are often not experiencing the same kinds of difficulties as peers who are less advantaged than they are. This negative form of self-organization influences repeated exclusion for students that fit into one or more of the categories listed above. Likely, the inexistence of learning opportunities to understand how to better meet these students' needs is unintentional. However, the problem is that as this continues to go unrecognized at this level, a gap between obtaining learning opportunities for teachers to develop efficacy, knowledge and skills for recognizing and leveraging the literacies of their students increases or, generally, stays the same.

In terms of social and emotional well-being, aspects of influence relating to the cultural and demographic relationships are apparent at the school levels where professional learning emerges from the collective self-organization of principal perspectives, teacher perspectives and student needs. For example, OES is undertaking a collective focus on vocabulary development, and educators in the urban school have had repeated opportunities to learn about integrating principals of attachment theory in their classrooms and their school. These foci reflect the recognition of uniquenesses on teacher learning at local levels emerging from students' needs nested in their particular systems. This research does not deeply explore the role of the principal in comparison to how it seeks to unpack and interpret the perspectives of teachers; yet a body of related empirical literature reflects how a sense of place influences principals' perspectives on local schooling needs (Abbate-Vaughn, 2004; Budge, 2006; Sider et al., 2017). The principal's role requires a more in-depth exploration of the range of factors influencing teachers' professional learning opportunities connected to social and emotional learning amongst early reading professional development.

Broadly, participants from all schools stress how social and emotional factors influence student learning. These factors relate to how children are affected by the changing pace and demands within society and the high number of students today who need help from their teachers and schools to nourish their social and emotional well-being. Without balancing social and emotional well-being, learning to read is an even more significant challenge. However, what is also apparent is that unfortunately, along with an overall sense of living in a rapid and disconnected society in comparison to prior generations, the local complexities at play are under-considered at times when thinking about student readiness within the confines of a curriculum privileging the school readiness of white upper middle class students (Perry, 2012). Students who go to school in urban centres and rural settings face sustaining and inequitably discordant opportunities to fail than peers where home life is considered more stable and literacy-enriched (Abbate-Vaughn, 2004; Barley & Beesley, 2007).

The power of complexity reduction flows from the social, emotional and behavioural diversities that students possess, and that appears to hinder their willingness or engagement with traditional academic expectations of learning to read. By considering how contextual factors at the student level are entrenched in social and emotional connections with the learning experience, reflects how local needs have to be responsibly and empathetically recognized and balanced within expectations for academic, and social and emotional development. Further, understanding the influential role of contextual factors indicates how important it is to create learning environments that are equitable and inclusive to students' social and emotional realities. Student needs vary, reflect their local contextual particularities and factors, and relate to the students' various connections, which need understanding and empathy from teachers to help make learning how to read a more meaningful and engaging event. A professional learning system that encompasses this kind of empathy shapes teachers' learning experiences based on their students' wide-ranging needs and provides learning experiences meaningfully engaging for their teachers.

The Learning Community

Priestley (2011) remarks how we live in a world of change and reform activity but that "the core of schooling has remained the same" (p. 2) and that this outdated mode of thinking is going to be detrimental to teaching and learning when it does not compassionately shift more towards the contextual elements influencing and representing students' and teachers' needs. Self-organized systems survive through changes experienced within and across the collectives of systems nested within systems (e.g., Morrison, 2008). Educational systems are human systems comprising a range of interacting collectives. In places, the findings in this research study articulate negativity from participants around reductions to professional learning and other

accessibility issues limiting experiences they yearn for or perceive as positive. For these
teachers, their needs mostly relate to their teaching context and teaching experiences, but their
learning experiences are dependent on assistance from learning systems able to ensure generating
and sustaining environments adaptive to teachers' and their students' learning needs.

Davis et al. (2012), in their complexity guided research with three school boards in
Western Canada, represent how a board (learning system) with strict centralized authority and
established hierarchies selected professional learning opportunities based on their value and
ability to have learning accountabilities measured. They demonstrate how a centralized learning
system reduces change opportunities by defining the structures of work and learning and how
teachers' roles and responsibilities are pre-established. This is not to imply that this kind of
system has ill intentions to those within but represents how the structures comprising the learning
system can constrict and limit learning experiences, how relationships form and the
environments where learning happens. In this type of relationship, teachers' perceptions of
decisions made higher up resemble, at times, confusion (e.g., questioning of motives and
directions) and a sense of being powerless to learn and change directions; a perception that BT1
and BT2 strongly embrace regarding the lack of control they perceive teachers generally have in
accessing professional learning. Permeating throughout the teacher-participant interviews with
CT1, CT2, BT1, and BT2 are their perceptions that professional learning opportunities have
steadily declined over the past few years, diminishing their opportunity to develop
professionally. OT1, OT2 and OP believe more needs to happen to plug rural teachers into face-
to-face professional learning networks. In terms of the relationship between systems nested
within systems Zellermayer & Margolin's (2005) complexity theory research relates how
individual identities develop through community participation and that communities are not

equal nor symmetrical. Learning communities have to be led by an empathetic leadership core group

> who organizes events and connects members of the community, as well as others who take
> on other leadership roles. These individuals form the core group of participants and
> actively take part in discussions in the public community forum. They often take on
> community projects, identify topics for the community to address, and move the
> community along its learning agenda. (p. 1277)

Zellermayer and Margolin indicate the importance of organizational capacity to make the learning experience an engaging and effective transition towards changes in understanding and practice when considering roles within a learning collaborative. In my research, participants appeared confused and surprised about the decline in their learning networks. Interestingly it is mainly the suburban and rural teachers who reflect perspectives of need, appreciation, and wistfulness towards teacher learning communities in their schools. BT1's and BT2's disappointment revolves around the loss of peer collaborative learning networks and opportunities to learn with expert early reading leaders. OT1's experience as a literacy rep for her school is a transformative experience for her, and she strongly wishes for more opportunities to learn with peers from other schools in the board. OT2's isolation is mostly in part to not having experiences for learning with other teachers who teach the same grade and does not feel this opportunity is on the horizon. When discussing changes to or severing of learning networks, perceptions of abandonment and loss of learning ensue. These perceptions indicate the influence of a learning system's capacity to have compassion for teachers' desires for learning and change. Opportunities for learning with peers from other schools is vital for these rural school teachers. For the suburban school, the two teacher participants gave fairly glowing opinions of their school (cycle) learning team. There is a perception of receiving professional nourishment through regularly sustained social learning experiences, with OT1, OT2, BT1 and BT2 sharing how they professionally benefit from these opportunities to meet their specific contextual needs.

Per the discussion at the start of this section was the notion that redundant human systems are by design organized in a harmful way to those closely connected within their learning system (Morrison, 2008; Priestly, 2011). This kind of learning system closes off the social nature of learning, restricting knowledge-on and knowledge-of-practice opportunities and projecting a position that knowledge resides within individuals and not across (Kelly, 2006). One case school board in Davis et al. (2012) crafted professional learning designs that severed collaboration opportunities beyond the school. This is a cognitivist, mechanistic approach to teacher learning and creates a barrier for extending experiences beyond present circumstances (Kelly, 2006). In Davis et al.'s research, this particular school board illustrates how a learning system organized to use special funds from the province to generate board and in-school experts for providing professional learning experiences based solely on school need limits collaboration and creates isolation, relating to the perceptions of exclusion OT1 and OT2 share. Though this approach ensures equality in opportunities for professional learning funding driven by local interests, it results in little within- and across-board school learning community (Davis et al., 2012).

A fragmented learning system represents an impersonal connection to schools and prevents fruitful and diverse learning relationships by operating through a model of efficiency: limiting the messiness of collaboration and keeping messages local and clear (Davis et al. 2012). In this kind of learning system, the learning collective represents a lost opportunity, and with it, the change opportunities that teachers, such as the participants in this research value, appreciate and crave. The teacher participants from OES illustrate teachers in a similar condition where they are at the mercy of learning primarily at their location and with a very restricted number of teachers. Collaboration provides one medium to overcome redundancy by welcoming and inquiring into differences in understanding, connecting through and to diversity, and allowing for new knowledge and practices to emerge.

All teachers in this research value their place in their learning communities, and though needs differ depending on teaching context, there are structural elements to the learning process that teachers from the three case schools identify as ways for meeting their professional early reading needs. Interestingly, though the two rural participants value their cycle partnership, because their school is isolated and has a small core group of teachers, they perceive that their school learning community's power to contribute to their professional learning is minimal. Often isolated rural schools create challenges for connecting teachers into learning communities, and it is harder to retain good teachers. These variables diminish opportunities for positive and sustaining initiatives (Budge, 2006; Hargreaves et al., 2015). OT1's and OT2's participation in across-board reading initiatives, when they do occur, appear to meet teachers' perceived early reading needs broadly. Participation in learning experiences focusing on setting up the reading environment and implementing some decoding and phonics-based instructional approaches has helped hone skills. Still, this does not appear to meet the sociocultural language-based needs these rural school teachers have. They suggest structuring professional learning authentic to the sociocultural, and relatedly (and subsequently) students' particular academic needs. For OT1 and OT2, when they have access to the broader professional community, they are appreciative, be it a mentor or time learning with groups of teachers from other schools. The three rural school participants remark that current modes of technological collaboration are not benefitting their professional learning. Research confirms the challenges of using technology as a useful professional learning tool for rural teachers, even though technology continues to emerge and make learning more accessible (Hunt-Barron et al., 2015). Teachers in rural schools appear to crave professional relationships that will motivate and influence their early reading instruction. It is essential to find more effective ways to utilize techno-collaborative tools in rural, isolated communities. Improving techno-collaborative learning experiences for rural teachers will provide the types of learning environments that Hargreaves et al. (2015) suggest is important when rural schools design professional learning.

Exemplifying an ideal learning system, Davis et al. describe how a school board that is a decentralized learning system conceptualizes its work as learning, learning involving risk-taking, change emerging in collaboration and articulates hierarchies in terms of responsibility rather than power and authority. In this system, collaboration encourages diverse knowledge and experiences and is a powerful way to develop learners and learn. While fragmented systems run the risk of having segregated learning communities and beliefs, and centralized systems can create learning opportunities that neglect diversity and contextual differences, decentralized systems embrace a collaborative learning environment where change is unpredictable to the degree that openness is valued. In my research, the case schools appear to be learning in a school board reflecting elements of all three kinds of learning systems Davis et al. (2012) discuss; inferring that positive learning experiences are happening but that at the same time, crucial forms of learning involving modes of collaboration are not.

Abbate-Vaughn (2005) describes how school demographics shape teacher perspectives on learning communities relating that teachers of mainly suburban middle-class children are more likely to view collaboration as something teachers design and implement as a team. It appears that this is the case for both BES teacher participants who find that school collaboration creates the space for a rich dialogue to emerge and influences their school's overall approach to reading instruction. For example, BT1 and BT2 perceive that board supported early reading professional learning acts as a catalyst to their reading instruction and is mostly responsible for shaping the rich communal conversations and shared language throughout the school. Their early reading professional learning sparks the development of practical learning opportunities within their school community. In contrast to suburban school teachers, Abbate-Vaughn finds that teachers in urban lower SES schools view their contribution to the school community as one part of the whole (learning in their silo is their way to more immediately connect and contribute to their students' success and relatedly their whole school's improvement). Neither of the urban school teacher participants in this study spoke about their teaching colleagues in too much detail, unlike participants from the two other case schools. Still, they share some positive perspectives of the two most recent early reading initiatives, but to different degrees. For example, at CES, CT2 is an expert mentor teacher of the reading programs and is frustrated with some of her colleagues' low levels of buy-in to these reading approaches. CT1 borrows elements of one reading program but does not dedicate too much time to its full-scale implementation. Further, though CES's principal mentions that networks are an ideal way to structure professional learning, it is not a structure that either teacher participant describes in detail. It will be interesting to explore this more deeply and understand how a large population of teachers in the same school teaching the same grades perceive their in-school collaborative learning experiences and how this influences and informs their reading instruction to leverage their students' literacy strengths.

A learning system that can question the constraints its style imposes upon those that are a part of it will be empathetic to enhancing teachers' and students' learning opportunities. The contextual variables representing the professional learning system display the learning system through its distribution of roles and responsibilities, the compassion and empathy to the needs of the individual and collective learners in the system and the environment in which learning is and should be occurring. Resonating in and out of this system discussion is the effect of learning structures on the individuals responsible for producing learning: teachers and students. Comprising a panoramic view of the learning, change and achievement process from the perspectives of the participants in three case schools within the school board illuminates how systems can generate learning opportunities that are locally and globally relevant with learning practices that emerge that are contextually considerate to the people at the core of the learning system.

Discussion Summary

Cilliers (2000) notes how emergence is not a random or statistical phenomenon, and that complexity thinking and research requires an exploration of the notions of the emergence of complex systems for an in-depth understanding of how natures emerge within particular systems. This understanding derives an intricate and personal contemplation of how systems interact and why it is necessary to unpack difference and diversity within systems. The depths and complexity of the within-case and cross-case analysis and findings are integral in understanding how the uniqueness of time and space influences individuals. Demographics, professional and personal experiences and circumstance, ease of opportunity and aspects of the teaching community, and a realm of other factors shape the opportunities, emotions and beliefs the six teacher participants hold towards early reading professional learning and practice. Pathways leading towards how teachers perceive the influence of elements of their professional learning on their practice and how it shapes their instruction for improving the early reading and literacy learning opportunities for their students situate the complexities of early reading professional learning and practice into a global snapshot within one school board. The cross-case analysis required a process of looking at these pathways in a holistic sense by weaving common findings together to portray a sense of how diverse schools within one school board share similarities in professional learning needs and interests. Together the school (within-case) and the board (cross-case) findings provide an inclusive interpretation of how context influences and shapes teachers' perspectives on their learning and practice. Via complex and interrelated variables and factors, a sense of challenges, needs, appreciations, inhibitors and influencers arise. Understanding perspectives in such a fashion provides the depth and richness that though unique to a specific board within a specific region of a specific province, can reverberate to alternative environments. By illuminating a way to conceptualize professional learning that considers teachers' readiness and needs, this research helps in clarifying challenges and complexities that exist within boards and also shines a light on the similarities to consider in conceptions of teachers' early reading professional learning that will positively influence students' reading achievement.

Without digging too deeply, it is probably easier to ascribe why too many students are not learning to read. For example, too many teachers have not developed a good knowledge and skill base for teaching students how to read. The solution to this could then be quickly answered by looking at the reading research and surveying early reading teachers to understand that they need to provide better code-based instruction and provide more learning opportunities for them to learn. However, taking a step back, there are so many interacting variables at play within teaching and learning systems that such a cut-and-dried solution is impractical. The solution presented very generally at the top of this paragraph reflects an impersonal and straightforward account of a human learning collective responsible for helping children learn and contribute to society. Though broad in scope and possibility, my research brings into this conversation on learning, practice and student reading achievement a conceptualization of a learning system that can connect more intimately with students' and teachers' needs. Very broadly, the contribution this research makes encompasses an understanding of the teachers' learning systems as empathetic and compassionate systems that influence the development of sensitive, adaptive and non-redundant learning environments. Chapter 6 concludes this study by suggesting theoretical, empirical and practical implications of this research for improving teachers' early reading professional learning and, subsequently, all students' early reading development.

Chapter 6 – Conclusion: Contributions and Implications

This complexity-guided research represents teachers' perceptions of their learning and teaching and the influences of these processes upon student reading development. Unpacked in this study are the influences of interactions within systems nested within systems and the psychological influences of contextual factors interacting to create tensions in and across teacher learning. A learning system such as a school board that is inclusively aware of the influences of varying contextual factors exudes a (pre)sense of being current and keeping current and includes an attachment to its past. Redundancy reduces opportunities for change, and systems aware of this can counter norms and stagnancies by honouring how individuals independently and collectively learn.

What emerges from this line of research and thinking is a conceptual framework that encourages school boards and provincial ministries of education to shift attention to how professional learning for their teachers can be more relatable to local and less-local needs. Briefly, this implies a research agenda that considers individual stages of teacher development, thoughtfully represents the local and universal needs of students in designs of professional learning, and by ensuring the learning system unites teachers within a school board in ways that relate to teachers' preferences for relevant, sustained, collaborative learning activities. Conceptualizing teachers' professional learning in this way represents inclusive and equitable education because it will connect teachers and students on the geographical, social, cultural and academic fringes.

It is important to iterate that this research does not discount the extensive research and conjecture about reading instruction and reading failure. On the contrary, I embrace the range of research on practical approaches to reading instruction, teacher learning, and knowledge, which influences student reading achievement. This body of research set my study in motion by helping illustrate a gap in early reading educational theory and informing how urgent it is to consider the influence of interactions in systems upon the teaching and learning processes. These findings add to early reading research in showing how teachers believe that their professional learning does not help them meet students' needs with reading difficulties. Therefore, this indicates that professional learning has to change to reverse this perspective. The following sections indicate new starting points for conceptualizing teacher learning and the empirical, theoretical, and practical implications of an emerging research program.

Theoretical Contribution

Theoretically, my research contributes to the complexity theory-based educational literature. The theoretical contribution adds to understanding the role of context on teachers' early reading professional learning and practice. A complexity theory lens has been applied to the fields of preservice teacher education (e.g., Cochran-Smith et al., 2014; Davis & Sumara 2012, Zellermayer & Margolin, 2005) and in considering characteristics of learning systems (Davis & Simmt, 2005; Davis et al., 2012), but complexity theory-guided research is lacking in the exploration of how a professional learning system influences teachers' early reading instruction as well as the depths of interrelating contextual factors upon teachers' early reading learning. This research has provided a lens to explore a significant problem too many children face and a problem that, for most who suffer it endure adverse consequences for the rest of their lives. Therefore, this research offers a way to theorize the nature of teachers' learning systems by putting students' and teachers' contextual influences under a comprehensive investigation to understand how the learning system can be reconsidered and regenerated to provide teachers better opportunities for meeting their students' early reading needs. The primary theoretical tool

204

that my research contributes is the conceptual framework for considering teacher learning, change and student early reading achievement.

This framework lays out the importance of deeply considering the role of context upon teachers' learning and how students and the context of the learning community interact in the teaching and learning process. The research undertaken in this doctoral book is highly personal. With teachers as the main participants, Seidman's (2006) in-depth interview process draws out teachers' lived experiences concerning their early reading instruction and learning. Within the interviews, I sought perceptions of context upon their teaching. Distilling the teacher participants' perspectives, the human nature of their teaching and learning results in the current conceptual framework depicted in Figure 9. This framework provides a roadmap for future research into teaching, student learning and the role of the learning system in making learning opportunities contextually sensitive for teachers and their students. Education systems are spaces and places of human interaction and development. By considering education systems to be based on equity and inclusion, this form of research will work towards a deeper understanding of how to address the contextual factors that need more recognition in professional learning designs.

Empirical Contribution to Research

My research study offers a broad lens to understand an ongoing problem and essential question: Why do so many students struggle to learn how to read? This research conceptualizes this issue as beyond one of teachers' lack of ability, knowledge and practical skills to teach students to read in the early years of schooling. Conducted through a complexity theory framework for designing research my study provides an in-depth demonstration of not only how teachers perceive their professional learning influences their instruction and subsequently their students' reading achievement, but it goes further to explore the contextual variables and factors

205

at play in teaching, learning and practice and these relationships to students' reading

development. My research seeks to go beyond the quantitative data showing how little teachers

know about early reading or about how well pre-ordinate early reading professional learning

designs in controlled settings can improve teacher practice and student reading achievement.

Seeking to open up these bodies of literature and focus on an in-depth exploration of teachers'

perceptions of how their learning experiences have influenced their instruction this research

provides a sense of the complex interactions of variables and factors (self, school, board and

beyond) at play in the life of the teacher and influencing theirs and their students' learning. The

findings in my research are derived from multiple stages of data analysis and demonstrate themes

particular to teachers in specific locations. For teachers in unique, non-suburban communities,

their students' demographic uniqueness heavily influences how they perceive their learning

opportunities influence their teaching.

However, after broadening the exploration towards the cross-case, my research

demonstrates how professional learning systems can conceptualize the general foundational

needs of their teachers (present and future). This need speaks to universal approaches to teaching

reading by demonstrating how teachers across the board do not perceive they can meet many of

their students who struggle to learn (e.g., identified learning disability, varying language and

literacy differences). Out of these similarities and differences emerges a new way to

conceptualize a teaching-learning system. In my research, I transition from within- to cross-case

analysis by adapting a method for representing causal networks. Borrowing from Miles et al.

(2014), I created contextual networks representing paths (and dead ends) of teachers from each

case school towards what they perceive as effective early reading professional learning and

effective in mediating reading development for students. This methodological borrowing underpins the cross-case findings and illuminates how three different schools in one board compare and contrast and that a deep understanding of this context can be derived by such a research method for drawing out highly contextualized details of participants that saturates a phenomenon under study and that creates meaningful ways for understanding and encouraging learning relationships.

In relation to an emerging research program, this study contributes to a methodological presence that will build upon the current conceptual framework. Setting the stage for future research on the factors at play and how they appear to influence the teaching and learning process requires more leg work to refine an understanding of how context (de)stabilizes learning. Expanding this style of research and qualitative case study approach to look across contextually diverse boards will help me refine this framework to better understand learning and change from varying perspectives. I envision a deeper understanding of systems thinking (school board and province) that considers the contextual variables and the contextual factors that can influence positive change by understanding what makes the relationships with learning more meaningful for teachers and consequently improves the opportunities for all students' early reading success. This information will generate a type of in-action-reaction early reading systems research with a range of schools in school boards and different provincial contexts. It will be by incorporating early-reading research, utilizing effective ways for developing teaching-learning collaboratives inclusive to different learning needs, and being guided in the different environmental and personal landscapes that situate teaching and learning, that a research program will emerge from the conceptual framework I have developed in this study. From the deep descriptive data collection, I believe that a contribution from forms of quantitative analyses will help expand current understandings about relationships between contextual factors and contextual variables

that will inform designs of qualitative data collection and analyses in research to improve
teacher-learning and student reading achievement.

Implications for Practice: Preservice and Inservice

In this section, I briefly present implications for practice that inform tentative guiding
principles for nudging changes to teacher knowledge and practice of early reading within
schools' current learning systems. These implications for practice connect to three levels:
preservice, inservice and at the provincial ministries and departments of education.

The preservice level. Incorporating a range of instructional strategies into practices
reflecting the rich and ethical tenets of whole reading and empathetic to the core code-based
reading evidence should be a part of pre-service teacher education. Connected to my research are
findings that indicate widespread low levels of teacher knowledge about code-based reading
(see, Joshi et al., 2009; Mathers et al., 2011; Stark et al., 2016.) At the preservice level, this
means enhancing content in courses that improve beginning teachers' knowledge in ways that
relate to effective practices. Along with ensuring code-based and whole language learning
opportunities pre-service literacy learning should also include opportunities to learn about the
different types of literacies students have and how to leverage diverse strengths in their future
classrooms.

The importance of learning a variety of early reading strategies in the beginning teaching
years reflects how influential this time is for long-lasting foundational skill development to take
root. Increase the focus on developing knowledge about learning to read and teaching practices
associated with including and teaching all students how to learn to read in bachelor of education
programs and providing preservice teachers practical experiences to develop knowledge and
practice, will yield more teachers who are prepared to effectively teach reading in their

beginning years. Not meeting teachers' learning needs in their beginning years will exacerbate problem areas across time (Shuck et al., 2018).

The inservice level. At the inservice level, and in the context of the school board in this research study, ensuring there are opportunities for early reading and language development within the new teacher induction program is one way for new teachers to develop lifelong core reading practices and social-literacy skills to help them empathetically meet their early readers' varying literacy needs. Seeking to develop ways to ensure responsive board learning structures for teachers to develop their current understanding of literacies and levels of early reading knowledge and practice is a reasonable way to increase board-wide reading achievement and decrease the reading achievement gap. A new teacher induction program attuned to the diversity of its teachers and students should pair teachers with whom they need, and in all likelihood, be considerate to teachers' readiness to provide effective early reading instruction. Shuck et al. (2018), and Butler and Schnellert (2012) emphasize how critical the right kind of mentorship program is in providing a teaching base to help teachers transition towards confidence and security in their self-efficacy. Research from Oliveira et al. (2019) extends this, highlighting how teacher coaches and mentors positively influence teacher reading knowledge and practice.

A valuable part of teachers' learning experience is developing relationships with peers in their professional learning communities. A system's empathy and compassion are catalysts to collaborative learning and represent components of a system that is sensitive to the particular needs of schools and relevant to its teachers (Hargreaves et al., 2015). The large body of research on school-based professional learning communities appears undeniably positive, relating teachers' positive perspectives of their learning communities upon their learning and in improving student achievement; and suggests that contemplating more contextually thoughtful professional learning designs will improve theirs and their students' learning (Abbate-Vaughn, 2004; Ford & Youngs, 2018; Kooy, 2015; Levine & Marcus, 2010; Vescio et al., 2008). Ensuring opportunities for teachers also to have ways for their local needs to be met as parts of these networks could anchor teacher early reading instruction more closely to their classroom and community contexts. Levine and Marcus (2010) indicate the need for intentionally designing collaborative learning communities and activities that can influence teacher change and student achievement. Teachers appreciate working with their peers in learning networks, and finding ways to sustain these kinds of programs appear to be essential for linking teachers across schools with less teacher diversity.

Rural schools can suffer from being self-isolated learning hubs. Isolation characterizes a fragmented learning system with tendencies for stifling opportunities for engaged learning and teacher development (Davis et al., 2012). Yuan et al. (2018) demonstrate how isolation inhibits growth, increases resentment and draws individuals inward. Technology's role is an important consideration when thinking about increasing community learning experiences and decreasing the perspective of being isolated. However, challenges in implementing technology intelligently for professional learning requires careful consideration (Clarke et al., 2017; Hunt-Barron et al., 2015). It is important to emphasize how much the rural participants in the current research study value face to face learning opportunities and that while increasing the capacity of technology to improve access to rural teachers' professional learning, technology should complement face-to-face learning. Web conferencing is something that rural teachers experience to be disengaging and unproductive. Better amalgamating technologically-based learning opportunities into teachers' distance, and online learning is a challenge that requires ongoing exploration. More consideration of the influence of technology upon professional learning will create fluid and conducive learning connections isolated teachers crave.

Influencing close relationships between school boards, schools, and the communities feeding into them are the school leaders' coordinated ability to connect schools into the community and set high expectations for student success (Barley & Beesley, 2007). Often this appears to involve fusing general and specific social, emotional and cultural characteristics of the community into the academic skill development of professional learning. Wright Jr. and Harris (2010) demonstrate the influence of board superintendents who recognized and were willing to find ways for their schools to adjust to students' cultural needs and, in doing so, demonstrated reductions in district achievement gaps. Schools inclusively responsive to their students' needs are likely associated with a board leadership empathetic to the central, unique needs students present, meeting these needs by helping teachers develop their knowledge and skills in these crucial areas. It makes sense to lay out the importance of establishing mutually positive relationships between teachers, schools and board. Ensuring transparency and committing to becoming aware of teachers' and students' specific needs is part of establishing a professional learning experience relevant to teachers and students (Wright Jr. & Harris, 2010).

School boards that ensure transparency will provide an essential personalized connection that some teachers want and are not getting. Considering teacher needs in terms of being led to professional learning opportunities or being informed of the directions for professional learning is also necessary for school boards to consider. For example, participants perceive that the school board was not transparent in helping teachers gain entry into the board's professional learning arc. Tones of isolation and animosity reflect these teachers' perceptions around being denied access to professional learning experiences they desire. Increased awareness of how teachers' social and emotional relations are affected by a system teachers perceive to be opaque would help to ease the tensions (e.g., confusion, animosity) about professional learning initiatives. In my research, teacher participants note that the professional learning communities they appreciate and have grown accustomed are diminishing. Suppose the board reduces opportunities for collaborative early reading professional learning. In that case, it also has to try to counter-balance itself so that new learning elements are equitably put into place to help teachers continue to develop a progressive range of early reading skills and knowledge, and that includes connections to their students with challenging social, emotional and academic needs (Abbate-Vaughn, 2004; Catalano et al., 2004; Chan, 2012; Forget-Dubois et al., 2009).

The provincial level. An inclusive, compassionate provincial ministry of education is another element that helps teachers obtain their professional learning needs. The importance of learning to read well is undeniable. Awareness of the need to have early reading instruction relate to the context and provide specific teaching skills to establish students' reading foundations needs more consideration at the provincial level. Clear guidelines will help provide direction and ensure teachers receive the experiences and the training they need along their career trajectories. For example, teachers coming out of their teacher education programs often carry with them the current attitudes and perspectives prevalent in their programs (Joshi et al., 2009; Moats & Lyons, 1989; Oliveira et al., 2019). These beginning teachers lack critical knowledge shown to influence student reading development, and these levels of knowledge often remain low even when teacher self-efficacy is high (Joshi et al., 2009; Stark et al., 2016; Wolf, 2018). Partly knowing what to teach, and what skills students should develop in the core areas of learning to read is a province's responsibility. For example, participants in my research shared their discomfort with a perceived lack of clarity about what their early reading instruction should entail in their curriculum. The overview of the reading instruction context in Québec presented in Chapter 2, notes the practical challenges to enacting the plan's ideological underpinnings. In Chapter 5, comments from teacher and school board participants express perceptions of imbalance between provincial accountability measures and how the curriculum expresses students will learn to read. Research exists to help teachers meet their students' early reading needs (code-based, positive and rich literacy-based environment). A curriculum that puts more explicit emphasis on what these skills are and how these stages develop will provide an environment conducive to early reading instruction for teachers and their students. Doing this should clear some confusion for beginning teachers and ensure consistency by helping all teachers frame their early reading instruction and allow new knowledge to emerge relating to students' localized needs.

A province can help make changes by bridging current research findings on early reading instruction with practical suggestions that open more learning opportunities for their early reading teachers. A transparent relationship between the province, board and schools will provide a chance for voices to be heard, needs to be voiced, and professional relationships established that are in the interest of all students regardless of the diversity and needs that schools across a province possess.

Limitations of this Study

This research was conducted with three schools in one school board and a particular province and is thus uniquely situated. English is one of two languages learned by students in their English Language School Board along with French. Language Arts comprises similar expectations to other provinces in terms of the number of hours across a school year, and so the students would be receiving the same official number of hours as peers across Canada. However, students in Québec English schools might not have the same amount of time to read in English in other subjects, which might influence reading experience and teachers' perceptions in ways that might not happen in other provinces. The perceptions provided in the three lengthy interviews with teachers provides a strong base of qualitative data to explore the phenomenon under study. However, there are going to be more contextual considerations that do not apply to teachers in this board and conducting similar research in another location will likely uncover important contextual factors when considering early reading professional learning designs.

I am interested in how the curriculum shapes teachers' perspectives or is perceived to influence how teachers approach and understand early reading in terms of their responsibility laid out in provincial curricula. The curriculum's role was alluded to in this research but was not explored too deeply regarding perspectives on professional learning and early reading instruction. It will be informative to understand how policy structures are perceived to influence learning opportunities and teaching change in providing more inclusive teaching practices to students who struggle to read. Research across provinces will yield a greater understanding of contextual factors in this level of the system and illuminate upon provincial structures empathetic in improving the learning environment.

Conclusion

This research provides fodder for future research in teachers' early reading professional learning by bringing context to the forefront of a research program. Looking at systems as a complex web of human relationships, the power of context upon teacher and student need is apparent. The implications for practice generated from these research findings provide some leading questions for future research. The conceptual framework provides a roadmap for empirical research that adapts early reading professional learning and makes it more contextually sensitive for teachers and creates more effective learning opportunities for their students. This research opens up the importance of understanding teachers' states and stages of their careers and adapting professional learning to this context and their students' specific contextual uniquenesses. The power of collaboration is undeniable, but for rural teachers, this is challenging, so one question is, "How can technology be utilized to create collaborative early reading learning opportunities rural teachers need and value?" It appears that in this case study, the school board collaborative learning opportunities in early reading are decreasing, so a question to ask is, "How can we create sustained professional learning communities in the times that these opportunities are diminishing?" Students' social, cultural, and emotional needs have to be understood and met as indicated primarily by urban and rural teachers. A question related to this is, "How can social and emotional skill development be merged with early reading professional learning so that students academically and socially connect to their classrooms, schools, and learning needs?" From the conversations with all participants and acknowledged in the literature, beginning teachers are not prepared to teach early reading, so a question to preservice teacher education programs is, "How can programs be better designed to prepare teacher candidates to meet the social and academic needs of their future students? Furthermore, how is literacy and language arts coursework going to prepare these students to be professionally ready to teach students who struggle to read?" These are just a few of the questions that arise out of my doctoral research study, and there are many more that have arisen out of this deep qualitative case study design. I believe that contextual considerations need to be at the forefront of systemic changes to create better and more equitable early reading learning opportunities for teachers and students.